PHYSICS RESEARCH AND TECHNOLOGY

FROM INFINITY TO INFINITY AND BEYOND

THE FIELD EVOLUTION EQUATIONS

PHYSICS RESEARCH AND TECHNOLOGY

Additional books in this series can be found on Nova's website
under the Series tab.

Additional e-books in this series can be found on Nova's website
under the e-book tab.

PHYSICS RESEARCH AND TECHNOLOGY

FROM INFINITY TO INFINITY AND BEYOND

THE FIELD EVOLUTION EQUATIONS

JANINA MARCIAK-KOZŁOWSKA
INSTITUTE OF ELECTRON TECHNOLOGY, WARSAW, POLAND

AND

MIROSLAW KOZLOWSKI
WARSAW UNIVERSITY, WARSAW, POLAND

New York

For permission to use material from this book please contact us:
Telephone 631-231-7269; Fax 631-231-8175
Web Site: http://www.novapublishers.com

Library of Congress Cataloging-in-Publication Data

ISBN: 978-1-63117-448-3

Published by Nova Science Publishers, Inc. † New York

CONTENTS

PREFACE

What is the origin of laws of physics ? Why do they have the form that they do, as opposed to a limitless number of other forms? When I was a student it was thought that the Big Bang was not just the origin of matter and energy, but of space and time too. Anything that came before the bang, or triggered it, was regarded as beyond the scope of science. Then in the 1980s, a number of cosmologists, notably James Hartle and Stephen Hawking, attempted to explain the Big Bang (including the origin of time) in terms of the laws of quantum gravity. For this type of explanation to succeed, the laws need to be transcendent: That is, they must exist in the absence of the universe—even in the absence of space and time.

Where, then, do these laws exist? Perhaps they occupy some abstract Platonic mathematical realm. Or maybe they live in the *multiverse*—a cosmic menagerie in which our universe is just one of many, each born in its own big bang. In this currently fashionable picture, the hot explosion that gave rise to our universe was not the ultimate origin of space and time, but just one spectacular event in an infinite and eternal space and time. The ultimate laws predate our universe because they already existed in the multiverse from which our universe sprang."

Furthermore, the laws are assumed to be fixed and immutable. Experiments support that assumption—no one has ever caught the laws of physics changing before his or her eyes—but we can only test the form of the laws to limited precision. Suppose the laws—even the ultimate high-temperature laws—evolve with time? What if they depend, perhaps in a subtle way, on the physical states of the universe, so that states and laws are coupled and can co-evolve? Such ideas are scientific heresy: The immutability of the laws has been at the foundation of physics since the time of Newton.

The laws of physics are normally cast as differential equations, which embed the concepts of real numbers and and of infinite and infinitesimal quantities as well as continuity of physical variables such as those of space and time.

The primordial Field F source of all laws of physics consists of five main fields

$$F=F_g+ F_{em} + F_s+ Fc$$

electromagnetic (F_{em}), gravitation (F_{grav}), strong (Fs) and consciousness field (Fc) Two of the fields: gravitation and electromagnetic are nonlocal and have the ranges $-\infty \rightarrow +\infty$, from *minus infinity* to *plus infinity and propagate with light velocity.* Considering that the radius of the present Universe is of the order of 10^{10} light years both electromagnetic and gravitation fields can be measured also beyond of the present Universe, which is finite. The range of stongfield is finite (10^{-15} m). The range of the consciousness field is still controversial. In this monograph we argue that consciousness field is additional degree of freedom and exists out of space- time, beyond infinity.

In the monograph we will describe the phenomena created by the electromagnetic and gravitation fields in the matter and in the spacetime free of matter-*void.*

Monograph *From infinity to infinity and beyond* consists of four chapters. In Part I *Electromagnetic Field* the master equation for the electromagnetic field interaction with matter is formulated and solved. It will be shown that the master equation is the generalized non-linear Klein-Gordon equations. The quanta of the Klein-Gordon field in matter are *the heatons with finite mass and velocity* $v=\alpha c$ where $\alpha=1/137$ and c is light velocity. In monograph phenomena in nanoscale as well as in femtoscale are described and solutions of Klein- Gordon equation are obtained.

In Part II, *Strong Field,* The Klein-Gordon equation is applied to the study of the phenomena in sub-nuclear scale. The existence of the free quarks is investigated. It was shown that in sub-nuclear scale the velocity propagation of the interaction $v=c$, $\alpha=1$.

In Part III *Field Equations for Life Science* moddified Klein-Gordond equation is applied to the study phenomena induced by absorption electromagnetic energy in biological materials: *cornea* and *cancer tumor.*

Part IV *Consciousness Field* is devoted to the study of consciousness field. The brain activity is investigated through the study of the spectrum of the brain waves. In Chapter IV the model for brain waves emission is presented. It is shown that the spectrum of brain waves can be described by Planck formula for black body radiation. As the result the temperature of the brain waves is calculated. Moreover it is shown that the brain waves spectrum have the same shape as the Cosmic Background radiation (but with different temperature). The model for brain waves emission is applied to the study of the “Langevin twins”.

PART I. ELECTROMAGNETIC FIELD

Chapter 1

INTRODUCTION

The source of electromagnetic field- laser aspires to be the next possible paradigm in Fundamental / Particle Physics. By its coherence, monochromaticity and field magnitude, it has been the lynchpin of novel spectroscopic methods of investigation that deepened our understanding of the atomic structure. However, it was inefficient to probe the subsequent strata formed by the nucleus, the nucleon or the vacuum. Neither the laser photon energy nor its electric field have been large enough to conceive decisive experiments beyond the atomic level. To reach the level where relevant nuclear and/or high energy physics investigations could be undertaken, large-scale laser infrastructures capable to deliver intensity in the ultra relativistic regime have been recently conceived based on the original concept introduced in 2002 (Tajima and Mourou) The first embodiment is ELI. It was launched under the aegis of the European Union community and built in Czech Republic, Hungary and Romania. It will yield the highest peak power and laser focused intensity. With its peak power of 100 PW, it represents the largest planned civilian laser project in the world. This gargantuan power will be obtained by delivering few kJ in 10 fs. Focusing this power over a micrometer spot size will yield intensities in the 1025 W/cm2 range, well into the ultrarelativistic regime. This extremely high peak intensity will correspond to the highest electric field, but also according to the pulse intensity-duration conjecture (Mourou and Tajima 2011) to the shortest pulse of high-energy particles and radiations, in the attosecond-zeptosecond regime.

Going beyond ELI is IZEST With ELI, the particle energy, radiation and field produced would reach the entry point for relevant Nuclear Physics, High Energy Physics or Vacuum The second initiative is promulgated by the International center on Zetta-Exawatt Science and Technology (IZEST) which was opened last year. It endeavors at the generation of exawatt-zettawat pulses produced by the delivery of greater than 10kJ in less than 10fs. It relies on already built large scale fusion lasers like the LMJ or NIF. To get around the grating damage threshold conundrum, a novel compression technique known as C3 (Cascaded Compression Conversion) was conceived It relies on the astute combination of the three compression techniques, CPA, OPCPA and Backward Raman Amplification(BRA). Based on plasma, C3 exhibits a much superior damage threshold (103-4) than CPA or OPCPA alone. It could potentially compress greater than 100kJ to the femtosecond regime paving the way to the exawatt-zettawatt regime and (*electromagnetic field* (laser) based particle physics.

Going to TeV Laser Accelerator with kJ Lasers When we try to reach for 100GeV and beyond (such as TeV) whose energies are sufficiently high for the frontier of high energy physics such as the search and study of Higgs boson, it is advantageous to employ kJ lasers). We realize that in order to reduce the required overall electric power needed to drive the accelerator, the average power of electricity to drive is proportional to the square root of the plasma density. This scaling also tells us that the laser energy per stage that is required to drive the laser accelerator is inversely proportional to the density to the power of 3/2. This means that in order to reduce the average power needed by 10 the cost of the accelerator by 10, we need to decrease the plasma density by 100, while we need to increase the individual laser energy by 1000. This is therefore preferred route towards the eventual high energy laser accelerator. It may be called the low density paradigm of laser acceleration. In addition to the above main advantage, it has a large number of superior performances as accelerator, such as less betatron radiation and consequentially reduced beam degradation, less stages and its therefore reduced beam degradation, smaller emittance degradation due to the jitters, etc. On the other hand, the expected elongation of the acceleration length is minor (as the optical connection between the stages are substantial and contributes to longer machine for higher densities) and in fact does not reflect in cost, as the cost is not much in creating gas tubes. In order to test and promote this paradigm, we need a 1-10kJ laser. IZEST equipped with the PETAL laser is ideally suited to first test this concept in this parameter regime. We plan to test toward 100GeV class experiments with or without staging. With sufficient amount of laser energy at PETAL and LMU, we regard that its extension can be TeV.

IZEST AND THE NONLUMINOSITY PARADIGM

We would like to point out that there are a class of fundamental physics questions that may be explored without having high luminosity. We suggest that the employment of MJ laser such as LMJ could reach TeV and even PeV Reaching this level of energies with even a small number of electrons has a significance, as we can test the validity of special theory of relativity by measuring the speed of light as a function of energy of the gamma photon in extreme high energies. This has been tried utilizing the arrivals of gamma photons (up to GeV) from Gamma Ray Bursts (GRBs), highest brightness of astrophysical objects thus farthest. Even though there appears to be some systematic delay of arrivals in higher end of gamma rays from GRBs, it is not clear if the delay is coming from the violation of Eintein's theory or due to the mechanism of accelerating high-energy particles in GRBs. Our laboratory acceleration can make a definitive test in par with the farthest edge GRB arrival, if we can make a precision measurement as we proposed. It is also possible to observe the synchrotron radiation spectrum from TeV and PeV electrons in these experiments. Synchrotron radiation is from the betatron oscillations of such high energy electrons in laser accelerationThere are a broader class of questions, however, in the extreme High Energy Cosmic Ray community that if there are highest energy cosmic rays in the range of 10^{21}eV. If so what are the objects that emit such and what are the mechanism for such. It is already clear to us that cosmic rays beyond 10^19eV that have been observed (up to 20^{20}eV) cannot be created by the well-know Fermi's acceleration. This is because the Fermi's acceleration assumes stochastic acceleration. When proton energy exceeds 10^{19}eV, the kick of protons in random directions would lose most of their energies into synchrotron radiations, even for protons. Thus it is incumbent for us to come up with a mechanism of extreme high energy acceleration. It may be possible to suggest that wakefield acceleration can provide a very compact prompt linear acceleration to extreme high energies that are required for this purpose In addition, we envision that the zeptosecond streaking we are considering using the colliding lasers and copropagating gamma photon can make time domain measurement of the vacuum. As particle pairs emerge from vacuum, we could test its mass dependence as a function of time. Such may be expected on quark pairs, as they are assumed to be clothed by gluons very heavily and their masses are dependent on how these gluon clothes are. In the naked vacuum state are quark mass much lighter than meson mass? This has been measured in high-energy collision events. Can we see this when they emerge from vacuum? Does its mass increases as it gets out of vacuum into a real particles? How? The proposed zeptosecond measurement may be able to answer this question. Furthermore, we might even see mass evolution of more mundane electrons and positrons.

HIGH-AVERAGE POWER LASERS: IZEST INTRODUCTION TO A NOVEL LASER ARCHITECTURE (ICAN EU PROJECT)

The low-density paradigm is beneficial for reducing the overall average power. However, we still need on the order of 10-100MW of electric power for lasers from the wallplug. In order to keep this number reasonable, it is also tantamount to have high efficiency of such drive lasers.

The ICAN (International Coherent Amplification Network) is a EU project (see the HP) addressing the question of high average power and high efficiency laser technology that is required for the driver of the laser accelerator based collider. ICAN has identified the fiber laser technology as the prime candidate for this driver. It has made significant technological leap so that we can begin to see the eventual outcome of this technology product now. The concept relies on the coherent phasing of a large multiplicity of Yb-doped fibers. The possibility to replace a single amplifying bulk material by a large number of fibers (greater than 10^4) increases enormously the cooling surface area. Fiber can also be pumped efficiently by laser diodes and offers a way towards superior laser wall plug to efficiency i.e >30%. Each fiber provides a single mode. Phased together they provide a way to reach single large pulse energy with great control. The noise analysis of single mode fiber shows that the main source of noise is thermal and at low frequencies (10Hz) making easy to control each fiber with electrooptic components. Nonlinear effects was also shown that they could be mitigated. Assuming 1mJ per fiber, CAN systems have the potential to deliver 10J/pulse, sub picosecond @10kHz or an average of 100kW with an efficiency surpassing 30% or 1000times what is demonstrated today. By combining a number of modules together the MW average could be easily been reached. Thus ICAN could be a fulcrum element of future laser-driven high energy accelerator. Let's note the project ICAN benefit from the invaluable experience of astronomers involved in the construction of the 42 m ESO telescope in Chilie. The telescope is composed of 1000 mirrors and 40000actuators activated a frequency of 2000 Hz. From the correction view point ICAN and the ETL are very close.

Vacuum Search by Intense Fields By employing copious coherent photons coming out from kJ-MJ lasers, we envision that the excitation of fields that have not manifested so far may become within our reach. The weakly coming dark particles escaped our detection so far, such as Dark Matter and Dark Energy, except for the astrophysical inferences. It remains a big mystery as of the property of these entities for now and laboratory experiments are badly needed. Homma et al. (2011; 2012) have introduced the degenerate four wave mixing technique to the possible detection of very weakly coupling fields by intense lasers. By virtue of the resonance of co-parallel laser pulses, we can enhance the gain by some 70 orders of magnitude if we apply kJ-MJ lasers over the charged particle collisions. Therefore, even weakly coupling Dark Matter candidates such as Axion-like Particles may be within the detectability and even more elusive Dark Energy candidates. Some candidates of these particles are believed to be very light. Therefore, it is best to employ light particles, in our case massless light itself, albeit with a huge quantity in coherence.

IZEST Exploring the Possibility to Produce Zettawatt

With the IZEST launch in 2011the theoretical possibility is coming into a more real project possibility, as PETAL and LMJ commit themselves toward fundamental research applications of kJ-MJ lasers. MJ systems are comprised of around 200beams of 10-20kJ beams. In order to realize coherent laser power at ZW, we need to have a technology that allow to compress and cohere a multiplicity of laser pulses into a single giant ultrashort pulse with peak power and intensity1000 times what is planned with ELI. Towards this goal the new concept (Mourou) was introduced, as mentioned earlier. When this level of large energy, ultrahigh intense laser pulses are generated, we are in the stage to explore not only the above laser acceleration in highest energy frontiers, but also in the nonlinear field search of vacuum, including QE vacuum and beyond. Also as mentioned, because of the intensity-pulse duration conjecture), this means the fastest possible optics. This will allow us to explore realtime view of physics in subatomic world, rather than the conventional energy spectrum views. For example, in the conventional technique we see the steady state, while in the time-domain approach utilizing ultrafast optics, we shall be able to see in principle such a phenomenon as the quarks (or electrons) coming out of vacuum first as bare elementary particles and later gradually wearing clothes to become real particles in a dynamical process.

Chapter 2

FUNDAMENTALS OF ELECTROMAGNETIC PULSES INTERACTION WITH MATTER

As early as 1956 M. Kac considered a particle moving on line at speed c, taking discrete steps of equal size, and undergoing collisions (reversals of direction) at random times, according to a Poisson process of intensity a. He showed that the expected position of the particle satisfies either of two difference equations, according to its initial direction. With correct scaling followed by a passage to the limit, the difference equations become a pair of first order partial differential equations (PDE). Differentiating those and adding them yields the hyperbolic diffusion equation

$$\frac{d^2u}{dt^2} + a\frac{du}{dt} = c\frac{d}{dx}\left(c\frac{du}{dx}\right). \tag{2.1}$$

This is an equation of hyperbolic type. If the lower term (in time) is dropped, it's just the one dimensional wave equation.

R. Hersh proposed the operator generalization of Eq. (2.1):

$$\frac{d}{dt}\left(\frac{dy}{dt}\right) + a\frac{dy}{dt} = A^2u \tag{2.2}$$

In equation (2.2) A is the generator of a group of linear operators acting on a linear space B. Instead of transition moving randomly to the left and right at speed c, the time evolution according to generators A and $*A$ is substituted.

The study and applications of the classical hyperbolic diffusion equation (2.1) covers the thermal processes the stock prices, astrophysics and heavy ion physics.

In this paragraph we will study the ultra-short thermal processes in the framework of the hyperbolic diffusion equation.

When an ultrafast electromagnetic pulse (e. g. femtosecond pulse) interacts with a metal surface, the excited electrons become the main carriers of the thermal energy. For a femtosecond thermal pulse, the duration of the pulse is of the same order as the electron relaxation time. In this case, the hyperbolicity of the thermal energy transfer plays an important role.

Radiation deposition of energy in materials is a fundamental phenomenon to laser processing. It converts radiation energy into material's internal energy, which initiates many thermal phenomena, such as heat pulse propagation, melting and evaporation. The operation of many laser techniques requires an accurate understanding and control of the energy deposition and transport processes. Recently, radiation deposition and the subsequent energy transport in metals have been investigated with picosecond and femtosecond resolutions. Results show that during high-power and short-pulse laser heating, free electrons can be heated to an effective temperature much higher than the lattice temperature, which in turn leads to both a much faster energy propagation process and a much smaller lattice-temperature rise than those predicted from the conventional radiation heating model. Corkum et al. found that this electron-lattice nonequilibrium heating mechanism can significantly increase the resistance of molybdenum and copper mirrors to thermal damage during high-power laser irradiation when the laser pulse duration is shorter than one nanosecond. Clemens et al. studied thermal transport in multilayer metals during picosecond laser heating. The measured temperature response in the first 20 ps was found to be different from predictions of the conventional Fourier model. Due to the relatively low temporal resolution of the experiment (~ 4 ps), however, it is difficult to determine whether this difference is the result of nonequilibrium laser heating or is due to other heat conduction mechanisms, such as non-Fourier heat conduction, or reflection and refraction of thermal waves at interfaces. Heat is conducted in solids through electrons and phonons. In metals, electrons dominate the heat conduction, while in insulators and semiconductors, phonons are the major heat carriers. Table 2.1 lists important features of the electrons and phonons.

Table 2.1. General Features of Heat Carriers [1]

	Free Electron	Phonon
Generation	ionization or excitation	lattice vibration
Propagation media	vacuum or media	media only
Statistics	Fermion	Boson
Dispersion	$E = \hbar^2 q^2/(2m)$	$E = E(q)$
Velocity (m·s^{-1})	~ 10^6	~ 10^3

The traditional thermal science, or macroscale heat transfer, employs phenomenological laws, such as Fourier's law, without considering the detailed motion of the heat carriers. Decreasing dimensions, however, have brought an increasing need for understanding the heat transfer processes from the microscopic point of view of the heat carriers. The response of the electron and phonon gases to the external perturbation initiated by laser irradiation can be described with the help of a memory function of the system. To that aim, let us consider the generalized Fourier law [1, 2]:

$$q(t) = -\int_{-\infty}^{t} K(t-t')\nabla T(t')dt', \tag{2.3}$$

where $q(t)$ is the density of a thermal energy flux, $T(t')$ is the temperature of electrons and $K(t - t')$ is a memory function for thermal processes. The density of thermal energy flux satisfies the following equation of heat conduction:

$$\frac{\partial}{\partial t}T(t) = \frac{1}{\rho c_v}\nabla^2 \int_{-\infty}^{t} K(t-t')T(t')dt', \tag{2.4}$$

where ρ is the density of charge carriers and c_v is the specific heat of electrons in a constant volume. We introduce the following equation for the memory function describing the Fermi gas of charge carriers:

$$K(t-t') = K_1 \lim_{t_0 \to 0} \delta(t - t' - t_0). \tag{2.5}$$

In this case, the electron has a very "short" memory due to thermal disturbances of the state of equilibrium. Combining Eqs. (2.5) and (2.4) we obtain

$$\frac{\partial}{\partial t}T(t) = \frac{1}{\rho c_v} K_1 \nabla^2 T. \tag{2.6}$$

Equation (2.6) has the form of the parabolic equation for heat conduction (PHC). Using this analogy, Eq. (2.6) may be transformed as follows:

$$\frac{\partial}{\partial t}T(t) = D_T \nabla^2 T. \tag{2.7}$$

where the heat diffusion coefficient D_T is defined as follows:

$$D_T = \frac{K_1}{\rho c_v}. \tag{2.8}$$

From Eq. (2.8), we obtain the relation between the memory function and the diffusion coefficient

$$K(t-t') = D_T \rho c_v \lim_{t_0 \to 0} \delta(t - t' - t_0). \tag{2.9}$$

In the case when the electron gas shows a "long" memory due to thermal disturbances, one obtains for memory function

$$K(t-t') = K_2 \tag{2.10}$$

When Eq. (2.10) is substituted to the Eq. (2.4) we obtain

$$\frac{\partial}{\partial t}T = \frac{K_2}{\rho c_v}\nabla^2 \int_{-\infty}^{t} T(t')dt, \tag{2.11}$$

Differentiating both sides of Eq. (2.11) with respect to *t*, we obtain

$$\frac{\partial^2 T}{\partial t^2} = \frac{K_2}{\rho c_v}\nabla^2 T. \tag{2.12}$$

Equation (2.12) is the hyperbolic wave equation describing thermal wave propagation in a charge carrier gas in a metal film. Using a well-known form of the wave equation,

$$\frac{1}{v^2}\frac{\partial^2 T}{\partial t^2} = \nabla^2 T. \tag{2.13}$$

and comparing Eqs. (2.12) and (2.13), we obtain the following form for the memory function:

$$K(t-t') = \rho c_v v^2 \tag{2.14}$$

v = finite, $v < \infty$.

As the third case, "intermediate memory" will be considered:

$$K(t-t') = \frac{K_3}{\tau}\exp\left[-\frac{(t-t')}{\tau}\right], \tag{2.15}$$

where τ is the relaxation time of thermal processes. Combining Eqs. (2.15) and (2.4) we obtain

$$c_v\frac{\partial^2 T}{\partial t^2} + \frac{c_v}{\tau}\frac{\partial T}{\partial t} = \frac{K_3}{\rho\tau}\nabla^2 T \tag{2.16}$$

and

$$K_3 = D_\tau c_v \rho. \tag{2.17}$$

Thus, finally,

$$\frac{\partial^2 T}{\partial t^2} + \frac{1}{\tau}\frac{\partial T}{\partial t} = \frac{D_\tau}{\tau}\nabla^2 T. \tag{2.18}$$

Equation (2.18) is the hyperbolic equation for heat conduction (HHC), in which the electron gas is treated as a Fermion gas. The diffusion coefficient D_T can be written in the form:

$$D_T = \frac{1}{3} v_F^3 \tau, \tag{2.19}$$

where v_F is the Fermi velocity for the electron gas in a semiconductor. Applying Eq. (2.19) we can transform the hyperbolic equation for heat conduction, Eq. (2.18), as follows:

$$\frac{\partial^2 T}{\partial t^2} + \frac{1}{\tau}\frac{\partial T}{\partial t} = \frac{1}{3} v_F^3 \nabla^2 T. \tag{2.20}$$

Let us denote the velocity of disturbance propagation in the electron gas as s:

$$s = \sqrt{\frac{1}{3}} v_F. \tag{2.21}$$

Using the definition of s, Eq. (2.20) may be written in the form

$$\frac{1}{s^2}\frac{\partial^2 T}{\partial t^2} + \frac{1}{\tau s^2}\frac{\partial T}{\partial t} = \nabla^2 T. \tag{2.22}$$

For the electron gas, treated as the Fermi gas, the velocity of sound propagation is described by the equation

$$v_s = \left(\frac{P_F^2}{3mm^*}\left(1 + F_0^S\right)\right)^{1/2}, \quad P_F = mv_F, \tag{2.23}$$

where m is the mass of a free (non-interacting) electron and m^* is the effective electron mass. Constant F_0^S represents the magnitude of carrier-carrier interaction in the Fermi gas. In the case of a very weak interaction, $m \to m^*$ and $F_0^S \to 0$, so according to Eq. (2.23),

$$v_S = \frac{mv_F}{\sqrt{3}m} = \sqrt{\frac{1}{3}} v_F. \tag{2.24}$$

To sum up, we can make a statement that for the case of weak electron-electron interaction, sound velocity $v_S = \sqrt{1/3} v_F$ and this velocity is equal to the velocity of thermal disturbance propagation s. From this we conclude that the hyperbolic equation for heat conduction Eq. (2.22), is identical as the equation for second sound propagation in the electron gas:

$$\frac{1}{v_S^2}\frac{\partial^2 T}{\partial t^2}+\frac{1}{\tau v_S^2}\frac{\partial T}{\partial t}=\nabla^2 T. \tag{2.25}$$

Using the definition expressed by Eq. (2.19) for the heat diffusion coefficient, Eq. (2.25) may be written in the form

$$\frac{1}{v_S^2}\frac{\partial^2 T}{\partial t^2}+\frac{1}{D_T}\frac{\partial T}{\partial t}=\nabla^2 T. \tag{2.26}$$

The mathematical analysis of Eq. (2.25) leads to the following conclusions:

1. In the case when $v_S^2 \to \infty$, τv_S^2 is finite, Eq. (2.26) transforms into the parabolic equation for heat diffusion:

$$\frac{1}{D_T}\frac{\partial T}{\partial t}=\nabla^2 T. \tag{2.27}$$

2. In the case when $\tau \to \infty$, v_S is finite, Eq. (2.26) transforms into the wave equation:

$$\frac{1}{v_S^2}\frac{\partial^2 T}{\partial t^2}=\nabla^2 T. \tag{2.28}$$

Equation (2.28) describes propagation of the thermal wave in the electron gas. From the point of view of theoretical physics, condition $v_S \to \infty$ violates the special theory of relativity. From this theory we know that there is a limited velocity of interaction propagation and this velocity $v_{lim} = c$, where c is the velocity of light in a vacuum.

Multiplying both sides of Eq. (2.26) by c^2, we obtain

$$\frac{c^2}{v_S^2}\frac{\partial^2 T}{\partial t^2}+\frac{c^2}{D_T}\frac{\partial T}{\partial t}=c^2\nabla^2 T, \tag{2.29}$$

Denoting $\beta = v_S/c$, Eq. (2.29) may be written in the form

$$\frac{1}{\beta^2}\frac{\partial^2 T}{\partial t^2}+\frac{1}{\widetilde{D}_T}\frac{\partial T}{\partial t}=c^2\nabla^2 T, \tag{2.30}$$

where $\widetilde{D}_T = \tau\beta^2$, $\beta < 1$. On the basis of the above considerations, we conclude that the heat conduction equation, which satisfies the special theory of relativity, acquires the form of the partial hyperbolic Eq. (2.30). The rejection of the first component in Eq. (2.30) violates the special theory of relativity.

Heat transport during fast laser heating of solids has become a very active research area due to the significant applications of short pulse lasers in the fabrication of sophisticated microstructures, synthesis of advanced materials, and measurements of thin film properties. Laser heating of metals involves the deposition of radiation energy on electrons, the energy exchange between electrons, and the lattice, and the propagation of energy through the media.

The theoretical predictions showed that under ultrafast excitation conditions the electrons in a metal can exist out of equilibrium with the lattice for times of the order of the electron energy relaxation time. Model calculations suggest that it should be possible to heat the electron gas to temperature T_e of up to several thousand degrees for a few picoseconds while keeping the lattice temperature T_l relatively cold. Observing the subsequent equilibration of the electronic system with the lattice allows one to directly study electron-phonon coupling under various conditions.

Several groups have undertaken investigations relating dynamics' changes in the optical constants (reflectivity, transmissivity) to relative changes in electronic temperature. But only recently, the direct measurement of electron temperature has been reported.

The temperature of hot electron gas in a thin gold film (l = 300 Å) was measured, and a reproducible and systematic deviation from a simple Fermi-Dirac (FD) distribution for short time $\Delta t \sim 0.4$ ps were obtained. The nascent electrons are the electrons created by the direct absorption of the photons prior to any scattering.

Tthe relaxation dynamics of the electron temperature with the hyperbolic heat transport equation (HHT), Eq.(2.26), can be investigated. Conventional laser heating processes which involve a relatively low-energy flux and long laser pulse have been successfully modeled in metal processing and in measuring thermal diffusivity of thin films. However, applicability of these models to short-pulse laser heating is questionable. As it is well known, the Anisimov model does not properly take into account the finite time for the nascent electrons to relax to the FD distribution. In the Anisimov model, the Fourier law for heat diffusion in the electron gas is assumed. However, the diffusion equation is valid only when relaxation time is zero, $\tau = 0$, and velocity of the thermalization is infinite, $v \to \infty$.

The effects of ultrafast heat transport can be observed in the results of front-pump back probe measurements. The results of these type of experiments can be summarized as follows. Firstly, the measured delays are much shorter than would be expected if heat were carried by the diffusion of electrons in equilibrium with the lattice (tens of picoseconds). This suggests that heat is transported via the electron gas alone, and that the electrons are out of equilibrium with the lattice on this time scale. Secondly, since the delay increases approximately linearly with the sample thickness, the heat transport velocity can be extracted, $v_h \cong 10^8\ \mathrm{cm \cdot s^{-1}} = 1\mu\mathrm{m \cdot ps^{-1}}$. This is of the same order of magnitude as the Fermi velocity of electrons in gold, $1.4\ \mu\mathrm{m \cdot ps^{-1}}$.

The effects of ultrafast heat transport can be observed in the results of front-pump back probe measurements. The results of these type of experiments can be summarized as follows: Firstly, the measured delays are much shorter than would be expected if the heat were carried by the diffusion of electrons in equilibrium with the lattice (tens of picoseconds). This suggests, that the heat is transported via the electron gas alone, and that the electrons are out of equilibrium with the lattice on this time scale. Secondly, since the delay increases approximately linearly with the sample thickness, the heat transport velocity can be extracted $v_h \cong 10^8\ \mathrm{cm \cdot s^{-1}} = 1\mu\mathrm{m \cdot ps^{-1}}$. This is of the same order of magnitude as the Fermi velocity of electrons in Au, $1.4\ \mu\mathrm{m \cdot ps^{-1}}$.

Since the heat moves at a velocity comparable to v_F - Fermi velocity of the electron gas, it is natural to question exactly how the transport takes place. Since those electrons which lie close to the Fermi surface are the principal contributors to transport, the heat-carrying electrons move at v_F. In the limit of lengths longer than the momentum relaxation length, λ, the random walk behavior is averaged and the electron motion is subject to a diffusion equation. Conversely, on a length scale shorter than λ, the electrons move ballistically with velocity close to v_F. The importance of the ballistic motion may be appreciated by considering the different hot-electron scattering lengths reported in the literature. The electron-electron scattering length in Au, λ_{ee} has been calculated]. They find that $\lambda_{ee} \sim (E - E_F)^2$ for electrons close to the Fermi level. For 2-eV electrons $\lambda_{ee} \approx$ 35nm increasing to 80 nm for 1-eV. The electron-phonon scattering length λ_{ep} is usually inferred from conductivity data. Using Drude relaxation times, λ_{ep} can be computed, $\lambda_{ep} \approx$ 42nm at 273 K. This is shorter than λ_{ee}, but of the same order of magnitude. Thus, we would expect that both electron-electron and electron-phonon scattering are important on this length scale. However, since conductivity experiments are steady state measurements, the contribution of phonon scattering in a femtosecond regime experiment such as pump-probe ultrafast lasers, is uncertain. In the usual electron-phonon coupling model, one describes the metal as two coupled subsystems, one for electrons and one for phonons. Each subsystem is in local equilibrium so the electrons are characterized by a FD distribution at temperature T_e and the phonon distribution is characterized by a Bose-Einstein distribution at the lattice temperature T_l. The coupling between the two systems occurs via the electron-phonon interaction. The time evolution of the energies in the two subsystems is given by the coupled parabolic differential equations (Fourier law). For ultrafast lasers, the duration of pump pulse is of the order of relaxation time in metals. In that case, the parabolic heat conduction equation is not valid and hyperbolic heat transport equation must be used (2.26)

$$\frac{1}{v_S^2}\frac{\partial^2 T}{\partial t^2} + \frac{1}{D_T}\frac{\partial T}{\partial t} = \nabla^2 T, \quad D_T = \tau v_S^2. \tag{2.31}$$

In Eq. (2.31), v_S is the thermal wave speed, τ is the relaxation time and D_T denotes the thermal diffusivity. In the following, Eq. (2.31) will be used to describe the heat transfer in the thin gold films.

To that aim, we define: T_e is the electron gas temperature and T_l is the lattice temperature. The governing equations for nonstationary heat transfer are

$$\frac{\partial T_e}{\partial t} = D_T \nabla^2 T - \frac{D_T}{v_S^2}\frac{\partial^2 T_e}{\partial t^2} - G(T_e - T_l), \qquad \frac{\partial T_l}{\partial t} = G(T_e - T_l). \tag{2.32}$$

where D_T is the thermal diffusivity, T_e is the electron temperature, T_e is the lattice temperature, and G is the electron-phonon coupling constant. In the following, we will assume that on subpicosecond scale the coupling between electron and lattice is weak and Eq. (2.32) can be replaced by the following equations (2.26):

$$\frac{\partial T_e}{\partial t} = D_T \nabla^2 T - \frac{D_T}{v_S^2}\frac{\partial^2 T_e}{\partial t^2}, \; T_l = \text{constant}. \tag{2.33}$$

Equation (2.33) describes nearly ballistic heat transport in a thin gold film irradiated by an ultrafast ($\Delta t < 1$ ps) laser beam. The solution of Eq. (2.33) for 1D is given by [1]:

$$T(x,t) = \frac{1}{v_S}\int dx' T(x',0) \left[\begin{array}{l} e^{-t/2\tau}\dfrac{1}{t_0}\Theta(t-t_0) + \\ e^{-t/2\tau}\dfrac{1}{2\tau}\left\{ \begin{array}{l} I_0\left(\dfrac{(t^2-t_0^2)^{1/2}}{2\tau}\right) \\ +\dfrac{t}{(t^2-t_0^2)^{1/2}} I_1\left(\dfrac{(t^2-t_0^2)^{1/2}}{2\tau}\right) \end{array} \right\}\Theta(t-t_0) \end{array} \right] \tag{2.34}$$

where v_s is the velocity of second sound, $t_0 = (x - x')/v_s$ and I_0 and I_1 are modified Bessel functions and $\Theta(t - t_0)$ denotes the Heaviside function. We are concerned with the solution to Eq. (2.34) for a nearly delta function temperature pulse generated by laser irradiation of the metal surface. The pulse transferred to the surface has the shape:

$$\Delta T_0 = \frac{\beta \rho_E}{C_V v_s \Delta t} \text{ for } 0 \le x \le v_s \Delta t,$$

$$\Delta T_0 = 0 \text{ for } x \ge v_s \Delta t \tag{2.35}$$

In Eq. (2.35), ρ_E denotes the heating pulse fluence, β is the efficiency of the absorption of energy in the solid, $C_V (T_e)$ is electronic heat capacity, and Δt is duration of the pulse. With $t = 0$ temperature profile described by Eq. (2.35) yields:

$$T(l,t) = \frac{1}{2}\Delta T_0 e^{-t/2\tau}\Theta(t-t_0)\Theta(t_0+\Delta t - t) \tag{2.36}$$

$$+\frac{\Delta t}{4\tau}\Delta T_0 e^{-t/2\tau}\left\{ I_0(z) + \frac{t}{2\tau}\frac{1}{z} I_1(z) \right\}\Theta(t-t_0),$$

where $z = (t^2 - t_0^2)^{1/2}/2\tau$ and $t = l/v_s$. The solution to Eq. (2.33), when there are reflecting boundaries, is the superposition of the temperature at l from the original temperature and from image heat source at $\pm 2nl$. This solution is:

$$T(l,t) = \sum_{i=0}^{\infty} \begin{array}{l} \Delta T_0 e^{-t/2\tau}\Theta(t-t_0)\Theta(t_0+\Delta t - t) + \\ \dfrac{\Delta t}{4\tau}\Delta T_0 e^{-t/2\tau}\left\{ I_0(z) + \dfrac{t}{2\tau}\dfrac{1}{z} I_1(z) \right\}\Theta(t-t_0), \end{array} \tag{2.37}$$

where $t_i = t_0$, $3t_0$, $5t_0$, $t_0 = l/v_0$. For gold, $C_V(T_e) = C_e(T_e) = \gamma T_e$, $\gamma = 71.5$ Jm^{-3} K^{-2} and Eq. (2.35) yields:

$$\Delta T_0 = \frac{1.4\times 10^5 \rho_E \beta}{v_s \Delta t T_e} \text{ for } 0 \le x \le v_s \Delta t$$
$$\Delta T_0 = 0 \text{ for } x \ge v_s \Delta t, \tag{2.38}$$

where ρ_E is measured in mJ · cm^{-2}, v_s in μm · ps^{-1}, and Δt in ps. For $T_e = 300$K:

$$\Delta T_0 = \frac{4.67\times 10^2 \rho_E \beta}{v_s \Delta t} \text{ for } 0 \le x \le v_s \Delta t$$
$$\Delta T_0 = 0 \text{ for } x \ge v_s \Delta t, \tag{2.39}$$

The model calculations (formulae 2.36 – 2.39) were applied to the description of the experimental results and a fairly good agreement of the theoretical calculations and experimental results was obtained.

In the early fifties it was shown by Dingle, Ward and Wilks and London, that a density fluctuation in a phonon gas would propagate as a thermal wave - a second sound wave - provided that "losses" from the wave were negligible. In one of their papers, Ward and Wilks indicated they would attempt to look for a second sound wave in sapphire crystals. No results of their experiments were published. Then, for nearly a decade, the subject of "thermal wave" lay dormant. Interest was revived in the sixties, primarily through the efforts of J. A. Krumhansl, R. A. Guyer and C. C. Ackerman. In the paper by Ackerman and Guyer the thermal wave in dielectric solids was experimentally and theoretically investigated. They found a value for the thermal wave velocity in LiF at a very low temperature $T \sim 1$ K, of $v_s \sim$ 100 – 300 ms^{-1}. In insulators and semiconductors phonons are the major heat carriers. In metals electrons dominate. For long thermal pulses, i.e., when the pulse duration, Δt, is larger than the relaxation time, τ, for thermal processes, $\Delta t >> \tau$, the heat transfer in metals is well described by Fourier diffusion equation. The advent of modern ultrafast lasers opens up the possibility investigating a new mechanism of thermal transport | the thermal wave in an electron gas heated by lasers. The effect of an ultrafast heat transport can be observed in the results of front pump back probe measurements. The results of this type of experiments can be summarized as follows. Firstly, the measured delays are much shorter than it would be expected if the heat were carried by the diffusion of electrons in equilibrium with the lattice (tens of picoseconds). This suggests that the heat is transported via the electron gas alone, and that the electrons are out of equilibrium with the lattice within this time scale. Secondly, since the delay increases approximately linearly with the sample thickness, the heat transport velocity can be determined, $v_h \sim 10^8$ cm s^{-1} = 1μm ps^{-1}. This is of the same order of magnitude as the Fermi velocity of electrons in Au, 1.4 μm ps^{-1}. Kozlowski et al [1] investigated the heat transport in a thin metal film (Au) with the help of the hyperbolic heat conduction equation. It was shown that when the memory of the hot electron gas in metals is taken into account, then the HHT is the dominant equation for heat transfer. The hyperbolic heat conduction equation for heat transfer in an electron gas has the form (2.26)

$$\frac{1}{\left(\frac{1}{3}v_F^2\right)}\frac{\partial^2 T}{\partial t^2}+\frac{1}{\tau\left(\frac{1}{3}v_F^2\right)}\frac{\partial T}{\partial t}=\nabla^2 T. \tag{2.40}$$

If we consider an infinite electron gas, then the Fermi velocity can be calculated

$$v_F \cong bc \tag{2.41}$$

In Eq. (2.41), c is the light velocity in vacuum and $b \sim 10^{-2}$. Considering Eq. (2.41), Eq. (2.40) can be written in a more elegant form:

$$\frac{1}{c^2}\frac{\partial^2 T}{\partial t^2}+\frac{1}{\tau c^2}\frac{\partial T}{\partial t}=\frac{b^2}{3}\nabla^2 T. \tag{2.42}$$

In order to derive the Fourier law from Eq. (2.42), we are forced to break the special theory of relativity and put in Eq. (2.42) $c \to \infty$; $\tau \to 0$. In addition, it can be demonstrated from HHT in a natural way, that in electron gas the heat propagation with velocity $v_h \sim v_F$ in the accordance with the results of the pump probe experiments.

Considering the importance of the thermal wave in future engineering applications and simultaneously the lack of the simple physics presentation of the thermal wave for engineering audience in the following we present the main results concerning the wave nature of heat transfer.

Hence, we discuss Eq. (2.42) in more detail. Firstly, we observe that the second derivative term dominates when:

$$c^2(\Delta t)^2 < c^2 \Delta t \tau \tag{2.43}$$

i.e., when $\Delta t < \tau$. This implies that for very short heat pulses we have a hyperbolic wave equation of the form:

$$\frac{1}{c^2}\frac{\partial^2 T}{\partial t^2}=\frac{b^2}{3}\nabla^2 T \tag{2.44}$$

and the velocity of the thermal wave is given by

$$v_{th} \sim \frac{1}{\sqrt{3}}\frac{c}{b},\quad b \sim 10^{-2}. \tag{2.45}$$

The velocity v_{th} in Eq. (2.45) is the velocity of the thermal wave in an infinite Fermi gas of electrons, which is free of all impurities. The thermal wave, which is described by the solution of Eq. (2.44), does not interact with the crystal lattice. It is the maximum value of the thermal wave obtainable in an infinite free electron gas. If we consider the opposite case to that in Eq. (2.43)

$$c^2(\Delta t)^2 > c^2 \Delta t \tau \tag{2.46}$$

i.e., when

$$\Delta t > \tau \tag{2.47}$$

then, one obtains from Eq. (2.42):

$$\frac{1}{\tau c^2}\frac{\partial T}{\partial t} = \frac{b^2}{3}\nabla^2 T. \tag{2.48}$$

Eq. (2.48) is the parabolic heat conduction equation – Fourier equation.

The value of the thermal wave velocity v_h is taken from paper [2.20]. Isotherms are presented as a function of the thin film thickness (length) l [μm] and the delay times. The mechanism of heat transfer on a nanometer scale, can be divided into three stages: a heat wave for $t \sim Lv_{th}^{-1}$, mixed heat transport for $Lv_{th}^{-1} < t < 3Lv_{th}^{-1}$ and diffusion for $t > 3Lv_{th}^{-1}$. The thermal wave moves in a manner described by the hyperbolic differential partial equation, $x = v_{th}\, t$. For $t < xv_{th}^{-1}$ the system is undisturbed by an external heat source (laser beam). For longer heat pulses the evidence of the thermal wave is gradually reduced - but the retardation of the thermal pulse is still evident.

If heat is released in a body of gas liquid or solid, a thermal flux transported by heat conduction appears. The pressure gradients associated with the thermal gradients set a gas or liquid in motion, so that additional energy transport occurs through convection. In particular, at sufficiently large energy releases, shock waves are formed in a gas or liquid which transport thermal energy at velocities larger that the speed of sound. Below the critical energy release, nearly pure thermal wave may propagate owing to heat conduction in a gas or liquid with other transport mechanisms being negligible. Solids metals provide an ideal test medium for the study of thermal waves, since they are practically incompressible at temperature below their melting point and the thermal wave pressures are small compared to the classic pressure (produced by repulsion of the atoms in the lattice) up to large energy releases. In accordance with this picture, the speed of sound in a metal is independent of temperature and given by $c_s = (E/\rho)^{1/2}$ where E is the elasticity modulus and ρ is the density.

Using the path-integral method developed in paper [2.31], the solution of the HHT can be obtained. It occurs, that the velocity of the thermal wave in medium is lower than the velocity of the initial thermal wave. The slowing of the thermal wave is caused by the scattering of heat carriers in medium. The scatterings also change the phase of the initial thermal wave.

In one-dimensional flow of heat in metals, the hyperbolic heat transport equation is given by (2.20).

$$\tau\frac{\partial^2 T}{\partial t^2} + \frac{\partial T}{\partial t} = D_T\frac{\partial^2 T}{\partial x^2}, \quad D_T = \frac{1}{3}v_F^3\tau, \tag{2.49}$$

where τ denotes the relaxation time, D_T is the diffusion coefficient and T is the temperature. Introducing the non-dimensional spatial coordinate $z = x / \lambda\!\!\!^{-}$, where $\lambda\!\!\!^{-} = \lambda / 2\pi$ denotes the reduced mean free path, Eq. (2.49) can be written in the form:

$$\frac{1}{v'^2}\frac{\partial^2 T}{\partial t^2} + \frac{2a}{v'^2}\frac{\partial T}{\partial t} = \frac{\partial^2 T}{\partial z^2}, \tag{2.50}$$

where

$$v' = \frac{v}{\lambda} \quad a = \frac{1}{2\tau} \tag{2.51}$$

In Eq. (2.51) v denotes the velocity of heat propagation [1], $v = (D/\tau)^{1/2}$.

In the paper by C. De Witt-Morette and See Kit Fong the path-integral solution of Eq. (2.50) was obtained. It was shown, that for the initial condition of the form:

$$T(z,0) = \Phi(z) \text{ an "arbitrary" function}$$

$$\frac{\partial T(z,t)}{\partial t}\Big|_{t=0} = 0 \tag{2.52}$$

the general solution of the Eq. (2.49) has the form:

$$\begin{aligned} T(z,t) &= \frac{1}{2}\left[\Phi(z,t) + \Phi(z,-t)\right]e^{-at} \\ &+ \frac{a}{2}e^{-at}\int_0^t d\eta\left[\Phi(z,\eta) + \Phi(z,-\eta)\right] \\ &+ \left[I_0(a(t^2-\eta^2)^{1/2}) + \frac{t}{(t^2-\eta^2)^{1/2}}I_1(a(t^2-\eta^2)^{1/2})\right] \end{aligned} \tag{2.53}$$

In Eq. (2.53), $I_0(x)$ and $I_1(x)$ denote the modified Bessel function of zero and first order respectively.

Let us consider the propagation of the initial thermal wave with velocity v', i.e.,

$$\Phi(z - v't) = \sin(z - v't) \tag{2.54}$$

In that case, the integral in (2.53) can be computed analytically, $\Phi(z, t) + \Phi(z, -t) = 2\sin z\cos(v't)$ and the integrals on the right-hand side of (2.53) can be done explicitly [2.31]; we obtain:

$$F(z,t) = e^{-at}\left[\frac{a}{w_1}\sin(w_1 t) + \cos(w_1 t)\right]\sin z,\ \ v' \geq a \tag{2.55}$$

and

$$F(z,t) = e^{-at}\left[\frac{a}{w_2}\sinh(w_2 t) + \cosh(w_2 t)\right]\sin z,\ \ v' < a \tag{2.56}$$

where $w_1 = (v'^2 - a^2)^{1/2}$ and $w_2 = (a^2 - v'^2)^{1/2}$.

In order to clarify the physical meaning of the solutions given by formulas (2.55) and (2.56), we observe that $v' = v/\lambda$ and w_1 and w_2 can be written as:

$$v_1 = \lambda w_1 = v\left(1 - \left(\frac{1}{2\tau\omega}\right)^2\right)^{1/2},\ \ 2\tau\omega > 1$$

$$v_2 = \lambda w_2 = v\left(\left(\frac{1}{2\tau\omega}\right)^2 - 1\right)^{1/2},\ \ 2\tau\omega < 1 \tag{2.57}$$

where ω denotes the pulsation of the initial thermal wave. From formula (2.57), it can be concluded that we can define the new effective thermal wave velocities v_1 and v_2. Considering formulas (2.56) and (2.57), we observe that the thermal wave with velocity v_2 is very quickly attenuated in time. It occurs that when $\omega^{-1} > 2\tau$, the scatterings of the heat carriers diminish the thermal wave.

It is interesting to observe that in the limit of a very short relaxation time, i.e., when $\tau \to 0$, $v_2 \to \infty$, because for $\tau \to 0$ Eq. (2.49) is the Fourier parabolic equation.

It can be concluded, that for $\omega^{-1} > 2\tau$, the Fourier equation is relevant equation for the description of the thermal phenomena in metals. For $\omega^{-1} > 2\tau$, the scatterings are slower than in the preceding case and attenuation of the thermal wave is weaker. In that case, $\tau \neq 0$ and v_1 is always finite:

$$v_1 = v\left(1 - \left(\frac{1}{2\tau\omega}\right)^2\right)^{1/2} < v \tag{2.58}$$

For $\tau \to 0$, i.e., for very rare scatterings $v_1 \to v$ and Eq. (2.49) is a nearly free thermal wave equation. For τ finite the $v_1 < v$ and thermal wave propagates in the medium with smaller velocity than the velocity of the initial thermal wave.

Considering the formula (2.55), one can define the change of the phase of the initial thermal wave β, i.e.:

$$\tan[\beta] = \frac{a}{w_1} = \frac{1}{2\tau\omega}\frac{1}{\sqrt{1-\frac{1}{4\tau^2\omega^2}}}, \quad 2\tau\omega > 1 \tag{2.59}$$

We conclude that the scatterings produce the change of the phase of the initial thermal wave. For $\tau \to \infty$ (very rare scatterings), $\tan[\beta] = 0$.

High-order wave equation for thermal transport phenomena

According that the complete Schrödinger equation has the form

$$i\hbar\frac{\partial\Psi}{\partial t} = -\frac{\hbar^2}{2m}\nabla^2\Psi + V\Psi \tag{2.60}$$

where V denotes the potential energy, one can obtain the new parabolic quantum heat transport going back to real time $t \to -2it$ and wave function $\Psi \to T$:

$$\frac{\partial T}{\partial t} = \frac{\hbar}{m}\nabla^2 T - \frac{2V}{\hbar}T \tag{2.61}$$

Equation (2.61) describer the quantum heat transport for $\Delta t > \tau$, where τ is the relaxation time. For heat transport initiated by ultrashort laser pulses, when $\Delta t > \tau$ one obtains the second order PDE for quantum thermal phenomena

$$\tau\frac{\partial^2 T}{\partial t^2} + \frac{\partial T}{\partial t} = \frac{\hbar}{m}\nabla^2 T - \frac{2V}{\hbar}T \tag{2.62}$$

Equation (2.62) can be written as

$$\frac{2V\tau}{\hbar}T + \tau\frac{\partial T}{\partial t} + \tau^2\frac{\partial^2 T}{\partial t^2} = \frac{\tau\hbar}{m}\nabla^2 T. \tag{2.63}$$

For distortionless thermal phenomena we obtain

$$V\tau \approx \frac{\hbar}{2}. \tag{2.64}$$

Equation (2.64) is Heisenberg uncertainty relation for thermal quantum phenomena. Substituting equation (2.64) to equation (2.63) we obtain the new form of quantum thermal equation

$$\left(1 + \tau\frac{\partial}{\partial t} + \tau^2\frac{\partial^2}{\partial t^2}\right)T = \frac{\tau\hbar}{m}\nabla^2 T. \tag{2.65}$$

It is obvious, from a dimensional analysis, that one can add the fourth term in equation (2.65), i.e.

$$\left(1+\tau\frac{\partial}{\partial t}+\tau^2\frac{\partial^2}{\partial t^2}+\tau^3\frac{\partial^3}{\partial t^3}\right)T=\frac{\tau\hbar}{m}\nabla^2 T. \tag{2.66}$$

When $V = 0$ equation (2.66) has the form

$$\left(\tau\frac{\partial}{\partial t}+\tau^2\frac{\partial^2}{\partial t^2}+\tau^3\frac{\partial^3}{\partial t^3}\right)T=\frac{\tau\hbar}{m}\nabla^2 T. \tag{2.67}$$

Let us write Eq. (2.67) in the form

$$\kappa\nabla^2 T=\varepsilon\frac{\partial^2 T}{\partial t^2}+\mu\frac{\partial T}{\partial t}+\mu_3\frac{\partial^3 T}{\partial t^3} \tag{2.68}$$

where

$$\kappa=\frac{\tau\hbar}{m},\ \varepsilon=\tau^2,\ \mu=\tau,\ \mu_3=\tau^3 \tag{2.69}$$

Equation (2.68) yields the characteristic polynomial equation

$$p(s,jk)=\mu_3 s^3+\varepsilon s^2+\mu s+\kappa k^2=0 \tag{2.70}$$

Equation (2.68) was investigated, for oscillating transport phenomena, by P.M. Ruiz. In one-dimensional case one obtains from Eq. (2.67)

$$\tau\frac{\partial T}{\partial t}+\tau^2\frac{\partial^2 T}{\partial t^2}+\tau^3\frac{\partial^3 T}{\partial t^3}=\frac{\tau\hbar}{m}\frac{\partial^2 T}{\partial x^2} \tag{2.71}$$

where

$$\tau\frac{\partial T}{\partial t}=\frac{\tau\hbar}{m}\frac{\partial^2 T}{\partial x^2} \tag{2.72}$$

is the diffusion equation.

Below we analyze the third-order wave equation

$$\tau^2\frac{\partial^3 T}{\partial t^3}=\frac{\hbar}{m}\frac{\partial^2 T}{\partial x^2} \tag{2.73}$$

in the case of thermal processes induced by attosecond laser pulses

$$\tau = \frac{\hbar}{mv^2}, \quad v = \alpha c, \tag{2.74}$$

and equation (2.73) can be rewritten as

$$\frac{\partial^2 T}{\partial x^2} = \beta \frac{\partial^3 T}{\partial t^3}, \quad \beta = \frac{\hbar}{mv^4}. \tag{2.75}$$

We seek a solution of equation (2.75) of the form

$$T(x,t) = Ae^{i(kx-\omega t)}. \tag{2.76}$$

Substituting equation (2.76) to Eq. (2.75) one obtains

$$(ik)^2 = \beta(-i\omega)^3. \tag{2.77}$$

This shows that equation (2.76) is the solution of the third-order PDE (2.75) i.e. Eq. (2.75) is the third-order wave equation if

$$\beta = \frac{(ik)^2}{(-i\omega)^3} = \frac{\hbar}{mv^2} \tag{2.78}$$

where v is the speed of propagation of thermal energy [2.33]. Substituting Eq. (2.78) to Eq. (2.76) one obtains

$$T(x,t) = Ae^{i\left[\frac{x}{\sqrt{2}\lambda}-\omega t\right]} e^{-\frac{1}{\sqrt{2}}\frac{x}{\lambda}} + Be^{-i\left[\frac{x}{\sqrt{2}\lambda}-\omega t\right]} e^{\frac{1}{\sqrt{2}}\frac{x}{\lambda}} \tag{2.79}$$

where λ is mean free path.

The second term in Eq. (2.79) tends to infinity for $x/\lambda >> 1$ and is to be omitted. The final solution of Eq. (2.76) has the form

$$T(x,t) = e^{-\frac{1}{\sqrt{2}}\frac{x}{\lambda}} Ae^{i\left[\frac{x}{\sqrt{2}\lambda}-\omega t\right]} \tag{2.80}$$

and describes the strongly damped thermal wave. It is interesting to observe that for electromagnetic interaction the third-order time derivative d^3x/dt^3 also describes the damping of the electron motion due to the self interaction of the charges.

REFERENCE

[1] M Kozłowski, J Marciak-Kozłowska, *Attoscience,* arXiv.0806.0165.

Chapter 3

QUANTUM EQUATION FOR FIELD PROPAGATION

Dynamical processes are commonly investigated using **electromagnetic field** pump-probe experiments with a pump pulse exciting the system of interest and a second probe pulse tracking is temporal evolution. As the time resolution attainable in such experiments depends on the temporal definition of the laser pulse, pulse compression to the attosecond domain is a recent promising development.

After the standards of time and space were defined the laws of classical physics relating such parameters as distance, time, velocity, temperature are assumed to be independent of accuracy with which these parameters can be measured. It should be noted that this assumption does not enter explicitly into the formulation of classical physics. It implies that together with the assumption of existence of an object and really independently of any measurements (in classical physics) it was tacitly assumed that *there was a possibility of an unlimited increase in accuracy of measurements.* Bearing in mind the "atomicity" of time i.e. considering the smallest time period, the Planck time, the above statement is obviously not true. Attosecond laser pulses we are at the limit of laser time resolution.

With attosecond laser pulses belong to a new Nano – World where size becomes comparable to atomic dimensions, where transport phenomena follow different laws from that in the macro world. This first stage of miniaturization, from 10^{-3} m to 10^{-6} m is over and the new one, from 10^{-6} m to 10^{-9} m just beginning. The Nano – World is a quantum world with all the predicable and non-predicable (yet) features.

In the sequent, we develop and solve the quantum relativistic transport equation for Nano – World transport phenomena where external forces exist.

There is an impressive amount of literature on hyperbolic heat transport in matter. In Chapter 2 we developed the new hyperbolic heat transport equation which generalizes the Fourier heat transport equation for the rapid thermal processes. The hyperbolic heat transport equation (HHT) for the fermionic system has be written in the form

$$\frac{1}{\left(\frac{1}{3}v_F^2\right)}\frac{\partial^2 T}{\partial t^2}+\frac{1}{\tau\left(\frac{1}{3}v_F^2\right)}\frac{\partial T}{\partial t}=\nabla^2 T, \tag{3.1}$$

where T denotes the temperature, τ the relaxation time for the thermal disturbance of the fermionic system, and v_F is the Fermi velocity.

In what follows we develop the new formulation of the HHT, considering the details of the two fermionic systems: electron gas in metals and the nucleon gas.

For the electron gas in metals, the Fermi energy has the form [1]

$$E_F^e = (3\pi)^2 \frac{n^{2/3}\hbar^2}{2m_e}, \tag{3.2}$$

where n denotes the density and m_e electron mass. Considering that

$$n^{-1/3} \sim a_B \sim \frac{\hbar^2}{me^2}, \tag{3.3}$$

and a_B = Bohr radius, one obtains

$$E_F^e \sim \frac{n^{2/3}\hbar^2}{2m_e} \sim \frac{\hbar^2}{ma^2} \sim \alpha^2 m_e c^2, \tag{3.4}$$

where c = light velocity and α = 1/137 is the fine-structure constant for electromagnetic interaction. For the Fermi momentum p_F we have

$$p_F^e \sim \frac{\hbar}{a_B} \sim \alpha m_e c, \tag{3.5}$$

and, for Fermi velocity v_F,

$$v_F^e \sim \frac{p_F}{m_e} \sim \alpha c. \tag{3.6}$$

Formula (3.6) gives the theoretical background for the result presented in Chapter 2. Comparing formulas (2.41) and (3.6) it occurs that $b = \alpha$. Considering formula (3.6), Eq. (2.42) can be written as

$$\frac{1}{c^2}\frac{\partial^2 T}{\partial t^2} + \frac{1}{c^2\tau}\frac{\partial T}{\partial t} = \frac{\alpha^2}{3}\nabla^2 T. \tag{3.7}$$

As seen from (3.7), the HHT equation is a relativistic equation, since it takes into account the finite velocity of light.

For the nucleon gas, Fermi energy equals

$$E_F^N = \frac{(9\pi)^{2/3}\hbar^2}{8mr_0^2}, \tag{3.8}$$

where m denotes the nucleon mass and r_0, which describes the range of strong interaction, is given by

$$r_0 = \frac{\hbar}{m_\pi c}, \tag{3.9}$$

wherein m_π is the pion mass. From formula (3.9), one obtains for the nucleon Fermi energy

$$E_F^N \sim \left(\frac{m_\pi}{m}\right)^2 mc^2. \tag{3.10}$$

In analogy to the Eq. (3.4), formula (3.10) can be written as

$$E_F^N \sim \alpha_s^2 mc^2, \tag{3.11}$$

where $\alpha_s = \frac{m_\pi}{m} \cong 0.15$ is the fine-structure constant for strong interactions. Analogously, we obtain the nucleon Fermi momentum

$$p_F^e \sim \frac{\hbar}{r_0} \sim \alpha_s mc \tag{3.12}$$

and the nucleon Fermi velocity

$$v_F^N \sim \frac{pF}{m} \sim \alpha_s c, \tag{3.13}$$

and HHT for nucleon gas can be written as

$$\frac{1}{c^2}\frac{\partial^2 T}{\partial t^2} + \frac{1}{c^2 \tau}\frac{\partial T}{\partial t} = \frac{\alpha_s^2}{3}\nabla^2 T. \tag{3.14}$$

In the following, the procedure for the discretization of temperature $T(\vec{r},t)$ in hot fermion gas will be developed. First of all, we introduce the reduced de Broglie wavelength

$$\begin{aligned} \lambda_B^e &= \frac{\hbar}{m_e v_h^e}, \quad v_h^e = \frac{1}{\sqrt{3}}\alpha c, \\ \lambda_B^N &= \frac{\hbar}{m v_h^N}, \quad v_h^N = \frac{1}{\sqrt{3}}\alpha_s c, \end{aligned} \tag{3.15}$$

and the mean free paths λ_e and λ_N

$$\lambda^e = v_h^e \tau^e, \; \lambda^N = v_h^N \tau^N. \tag{3.16}$$

In view of formulas (3.15) and (3.16), we obtain the HHC for electron and nucleon gases

$$\frac{\lambda_B^e}{v_h^e}\frac{\partial^2 T}{\partial t^2} + \frac{\lambda_B^e}{\lambda^e}\frac{\partial T}{\partial t} = \frac{\hbar}{m_e}\nabla^2 T^e, \tag{3.17}$$

$$\frac{\lambda_B^N}{v_h^N}\frac{\partial^2 T}{\partial t^2} + \frac{\lambda_B^N}{\lambda^N}\frac{\partial T}{\partial t} = \frac{\hbar}{m}\nabla^2 T^N. \tag{3.18}$$

Equations (3.17) and (3.18) are the hyperbolic partial differential equations which are the master equations for heat propagation in Fermi electron and nucleon gases. In the following, we will study the quantum limit of heat transport in the fermionic systems. We define the quantum heat transport limit as follows:

$$\lambda^e = \lambda_B^e, \; \lambda^N = \lambda_B^N. \tag{3.19}$$

In that case, Eqs. (3.17) and (3.18) have the form

$$\tau^e \frac{\partial^2 T^e}{\partial t^2} + \frac{\partial T^e}{\partial t} = \frac{\hbar}{m_e}\nabla^2 T^e, \tag{3.20}$$

$$\tau^N \frac{\partial^2 T^N}{\partial t^2} + \frac{\partial T^N}{\partial t} = \frac{\hbar}{m}\nabla^2 T^N, \tag{3.21}$$

where

$$\tau^e = \frac{\hbar}{m_e\left(v_h^e\right)^2}, \; \tau^N = \frac{\hbar}{m\left(v_h^N\right)^2}. \tag{3.22}$$

Equations (3.20) and (3.21) define the master equation for quantum heat transport (QHT). Having the relaxation times τ^e and τ^N, one can define the “pulsations” ω_h^e and ω_h^N

$$\omega_h^e = (\tau^e)^{-1}, \; \omega_h^N = (\tau^N)^{-1}, \tag{3.23}$$

or

$$\omega_h^e = \frac{m_e\left(v_h^e\right)^2}{\hbar}, \quad \omega_h^N = \frac{m\left(v_h^N\right)^2}{\hbar},$$

i.e.,

$$\omega_h^e\hbar = m_e\left(v_h^e\right)^2 = \frac{m_e\alpha^2}{3}c^2,$$
$$\omega_h^N\hbar = m\left(v_h^N\right)^2 = \frac{m\alpha_s^2}{3}c^2. \tag{3.24}$$

The formulas (3.24) define the Planck-Einstein relation for heat quanta E_h^e and E_h^N

$$E_h^e = \omega_h^e\hbar = m_e\left(v_h^e\right)^2,$$
$$E_h^N = \omega_h^N\hbar = m_N\left(v_h^N\right)^2. \tag{3.25}$$

The heat quantum with energy $E_h = \hbar\omega$ can be named the *heaton*, in complete analogy to the *phonon*, *magnon*, *roton*, etc. For τ^e, $\tau^N \to 0$, Eqs. (3.20) and (3.24) are the Fourier equations with quantum diffusion coefficients D^e and D^N

$$\frac{\partial T^e}{\partial t} = D^e\nabla^2 T^e, \quad D^e = \frac{\hbar}{m_e}, \tag{3.26}$$

$$\frac{\partial T^N}{\partial t} = D^N\nabla^2 T^N, \quad D^N = \frac{\hbar}{m}. \tag{3.27}$$

The quantum diffusion coefficients D^e and D^N were introduced for the first time by E. Nelson.

For finite τ^e and τ^N, for $\Delta t < \tau^e$, $\Delta t < \tau^N$, Eqs. (3.20) and (3.21) can be written as

$$\frac{1}{(v_h^e)^2}\frac{\partial^2 T^e}{\partial t^2} = \nabla^2 T^e, \tag{3.28}$$

$$\frac{1}{(v_h^N)^2}\frac{\partial^2 T^N}{\partial t^2} = \nabla^2 T^N. \tag{3.29}$$

Equations (3.28) and (3.29) are the wave equations for quantum heat transport (QHT). For $\Delta t > \tau$, one obtains the Fourier equations (3.26) and (3.27).

In what follows, the dimensionless form of the QHT will be used. Introducing the reduced time t' and reduced length x',

$$t' = t/\tau, \quad x' = \frac{x}{v_h \tau}, \tag{3.30}$$

one obtains, for QHT,

$$\frac{\partial^2 T^e}{\partial t^2} + \frac{\partial T^e}{\partial t} = \nabla^2 T^e, \tag{3.31}$$

$$\frac{\partial^2 T^N}{\partial t^2} + \frac{\partial T^N}{\partial t} = \nabla^2 T^N. \tag{3.32}$$

and, for QFT,

$$\frac{\partial T^e}{\partial t} = \nabla^2 T^e, \tag{3.33}$$

$$\frac{\partial T^N}{\partial t} = \nabla^2 T^N. \tag{3.34}$$

Electromagnetic phenomena in vacuum are characterized by two three dimensional vector fields, the electric and magnetic fields $E(x,t)$ and $B(x,t)$ which are subject to Maxwell's equation and which can also be thought of as the classical limit of the quantum mechanical description in terms of photons. The photon mass is ordinarily assumed to be exactly zero in Maxwell's electromagnetic field theory, which is based on gauge invariance. If gauge invariance is abandoned, a mass term can be added to the Lagrangian density for the electromagnetic field in a unique way:

$$L = -\frac{1}{4\mu_0} F_{\mu\nu} F^{\mu\nu} - j_\mu A_\mu + \frac{\mu_\gamma^2}{2\mu_0} A_\mu A^\mu \tag{3.35}$$

where μ_γ^{-1} is a characteristic length associated with photon rest mass, A_μ and j_μ are the four - dimensional vector potential $\left(\vec{A},\ i\phi/c\right)$ and four – dimensional vector current density $\left(\vec{J},\ ic_\rho\right)$ with ϕ and $\vec{A}$ denoting the scalar and vector potential and ρ, $\vec{J}$ are the charge and current density, respectively, μ_0 is the permeability constant of free space and $F_{\mu\nu}$ is the antisymmetric field strength tensor. It is connected to the vector potential through

$$F_{\mu\nu} = \frac{\partial A_\nu}{\partial x_\mu} - \frac{\partial A_\mu}{\partial x_\nu} \tag{3.36}$$

The variation of Lagrangian density with respect to A_μ yields the Proca equation (Proca 1930)

$$\frac{\partial F_{\mu\nu}}{\partial x_\nu} + \mu_\gamma^2 A_\mu = \mu_0 J_\mu \tag{3.37}$$

Substituting Eq.(3.36) into (3.37) we obtain the wave equation of the Proca field

$$\left(\nabla^2 - \frac{\partial^2}{\partial (ct)^2} - \mu_\gamma^2\right) A_\mu = -\mu_0 J \tag{3.38}$$

$$(\Box - \mu_\gamma^2) A_\mu = -\mu_0 J \tag{3.39}$$

In free space The *Proca* equation (3.39) reduces to (3.40), for a vector electromagnetic potential of $\boldsymbol{A_\mu}$.

$$\begin{aligned} (\Box + \mu_\gamma^2) A_\mu &= 0, \\ \Box &= \frac{1}{c^2}\frac{\partial^2}{\partial t^2} - \frac{\partial^2}{\partial x^2} \end{aligned} \tag{3.40}$$

which is essentially the Klein – Gordon equation for massive photons. The parameter μ_γ can be interpreted as the photon rest mass m_γ with

$$m_\gamma = \frac{\mu_\gamma \hbar}{c}. \tag{3.41}$$

With this interpretation the characteristic scaling length μ_γ^{-1} becomes the reduced Compton wavelength of the photon interaction. An additional point is that static electric and magnetic fields would exhibit exponential dumping governed by the term $\exp\left(-\mu_\gamma^{-1} r\right)$ is the photon is massive instead of massless.

SPECIAL RELATIVITY WITH NONZERO PHOTON MASS

It is well-known that the electromagnetic constant c in Maxwell theory of electromagnetic waves propagating in vacuum and special relativity was developed as a consequence of the constancy of the speed of light. However, one of the prediction of massive photon electromagnetic theory is that there will be dispersion of the velocity of massive photon in vacuum.

The plane wave solution of the Proca equations without current is $A^{\nu} \sim \exp\left(ik^{\mu}x_{\mu}\right)$, where the wave vector $k^{\mu} = \left(\omega, \vec{k}\right)$ satisfies the relationship

$$k^2c^2 = \omega^2 - \mu_{\gamma}^2c^2 \tag{3.42}$$

As can be shown that

$$v_g \text{ (group velocity)} = c\left(1 - \frac{\mu_{\gamma}^2c^2}{2\omega^2}\right) \tag{3.43}$$

and

$$v_g = 0 \quad \text{for} \quad \omega = \mu_{\gamma}c$$

namely the massive waves do not propagate. When $\omega < \mu_{\gamma}c$, k becomes an imaginary quantity and the amplitude of a free massive wave would, therefore, be attenuated exponentially. Only when $\omega > \mu_{\gamma}c$ can the waves propagate in vacuum unattenuated. In the limit $\omega \to \infty$, the group velocity will approach the constant c for all phenomena.

A nonzero photon mass implies that the speed of light is not unique constant but is a function of frequency. In fact, the assumption of the constancy of speed of light is not necessary for the validity of special relativity (Szymacha), i.e. special relativity can instead be based on the existence of a unique limiting speed c to which speeds of all bodies tend when their energy becomes much larger then their mass. Then, the velocity that enters in the Lorentz transformation would simply be this limiting speed, not the speed of light

It is quite interesting that the *Proca* type equation can be obtained for thermal phenomena. In the following starting with the hyperbolic heat diffusion equation the *Proca* equation for thermal processes will be developed and solved.

The relativistic hyperbolic transport equation can be written as

$$\frac{1}{v^2}\frac{\partial^2 T}{\partial t^2} + \frac{m_0\gamma}{\hbar}\frac{\partial T}{\partial t} = \nabla^2 T. \tag{3.44}$$

In equation (3.44) v is the velocity of heat waves, m_0 is the mass of heat carrier and γ – the Lorentz factor, $\gamma = \left(1 - \frac{v^2}{c^2}\right)^{-1/2}$. As was shown in paper [1] the heat energy (*heaton temperature*) T_h can be defined as follows:

$$T_h = m_0 \gamma v^2. \tag{3.45}$$

Considering that v, the thermal wave velocity equals:

$$v = \alpha c \tag{3.46}$$

where α is the coupling constant for the interactions which generate the *thermal wave* (α = 1/137 and α = 0.15 for electromagnetic and strong forces respectively). The *heaton temperature* is equal to

$$T_h = \frac{m_0 \alpha^2 c^2}{\sqrt{1-\alpha^2}}. \tag{3.47}$$

Based on equation (3.47) one concludes that the *heaton temperature* is a linear function of the mass m_0 of the heat carrier. It is interesting to observe that the proportionality of T_h and the heat carrier mass m_0 was observed for the first time in ultrahigh energy heavy ion reactions measured at CERN M. Kozlowski et al [1] has shown that the temperature of pions, kaons and protons produced in Pb+Pb, S+S reactions are proportional to the mass of particles. Recently, at Rutherford Appleton Laboratory (RAL), the VULCAN LASER was used to produce the elementary particles: electrons and pions.

M. Kozlowski et al developed wave equation

$$\frac{1}{v^2}\frac{\partial^2 T}{\partial t^2} + \frac{m}{\hbar}\frac{\partial T}{\partial t} + \frac{2Vm}{\hbar^2}T - \nabla^2 T = 0. \tag{3.48}$$

The relativistic generalization of equation (3.48) is quite obvious:

$$\frac{1}{v^2}\frac{\partial^2 T}{\partial t^2} + \frac{m_0\gamma}{\hbar}\frac{\partial T}{\partial t} + \frac{2Vm_0\gamma}{\hbar^2}T - \nabla^2 T = 0. \tag{3.49}$$

It is worthwhile noting that in order to obtain a non-relativistic equation we put $\gamma = 1$.

When the external force is present $F(x,t)$ the forced damped transport is obtained instead of equation (3.49) (in one dimensional case):

$$\frac{1}{v^2}\frac{\partial^2 T}{\partial t^2} + \frac{m_0\gamma}{\hbar}\frac{\partial T}{\partial t} + \frac{2Vm_0\gamma}{\hbar^2}T - \frac{\partial^2 T}{\partial x^2} = F(x,t). \tag{3.50}$$

The hyperbolic relativistic quantum heat transport equation, (3.50), describes the forced motion of heat carriers which undergo scattering ($\frac{m_0\gamma}{\hbar}\frac{\partial T}{\partial t}$ term) and are influenced by the potential term ($\frac{2Vm_o\gamma}{\hbar^2}T$).

Equation (3.50) can be written as

$$\left(\Diamond + \frac{2Vm_0\gamma}{\hbar^2}\right)T + \frac{m_0\gamma}{\hbar}\frac{\partial T}{\partial t} = F(x,t),. \qquad \Diamond = \frac{1}{v^2}\frac{\partial^2}{\partial t^2} - \frac{\partial^2}{\partial x^2}. \tag{3.51}$$

We seek the solution of equation (3.51) in the form

$$T(x,t) = e^{-t/2\tau}u(x,t) \tag{3.52}$$

where $\tau = \hbar/(mv^2)$ is the relaxation time. After substituting equation (3.52) in equation (3.51) we obtain a new equation

$$\left(\Diamond + q^2\right)u(x,t) = e^{t/2\tau}F(x,t) \tag{3.53}$$

and

$$q^2 = \frac{2Vm}{\hbar^2} - \left(\frac{mv}{2\hbar}\right)^2 \tag{3.54}$$

$$m = m_0\gamma \tag{3.55}$$

In free space i.e. when $F(x,t) \to 0$ equation (3.53) reduces to

$$\left(\Diamond + q^2\right)u(x,t) = 0 \tag{3.56}$$

which is essentially the free *Proca* equation, compare equation (3.40).

The *Proca* equation describes the interaction of the electromagnetic pulse with the matter. As was shown by Kozlowski et al., the quantization of the temperature field leads to the *heatons* – quanta of thermal energy with a mass $m_h = \hbar/\tau v_h^2$, where τ is the relaxation time and v_h is the finite velocity for heat propagation. For $v_h \to \infty$, i.e. for $c \to \infty$, $m_o \to 0$. it can be concluded that in non-relativistic approximation (c = infinite) the *Proca* equation is the diffusion equation for massless photons and heatons.

SOLUTION OF THE PROCA THERMAL EQUATION

For the initial *Cauchy* condition:

$$u(x,0) = f(x), \qquad u_t(x,0) = g(x) \tag{3.57}$$

the solution of the *Proca* equation has the form (for $q > 0$) [3]

$$\begin{aligned} u(x,t) &= \frac{f(x-vt)+f(x+vt)}{2} \\ &+ \frac{1}{2v}\int_{x-vt}^{x+vt} g(\varsigma) J_0\left[q\sqrt{v^2t^2-(x-\varsigma)^2}\right] d\varsigma \\ &- \frac{\sqrt{q}vt}{2}\int_{x-vt}^{x+vt} f(\varsigma)\frac{J_1\left[q\sqrt{v^2t^2-(x-\varsigma)^2}\right]}{\sqrt{v^2t^2-(x-\varsigma)^2}} d\varsigma \\ &+ \frac{1}{2v}\int_0^t \int_{x-v(t-t')}^{x+v(t-t')} G(\varsigma,t') J_0\left[q\sqrt{v^2(t-t')^2-(x-\varsigma)^2}\right] dt' d\varsigma. \end{aligned} \tag{3.58}$$

where $G = e^{t/2\tau} F(x,t)$.

When $q < 0$ solution of *Proca* equation has the form [1,3]:

$$\begin{aligned} u(x,t) &= \frac{f(x-vt)+f(x+vt)}{2} \\ &+ \frac{1}{2v}\int_{x-vt}^{x+vt} g(\varsigma) I_0\left[-q\sqrt{v^2t^2-(x-\varsigma)^2}\right] d\varsigma \\ &- \frac{\sqrt{-q}vt}{2}\int_{x-vt}^{x+vt} f(\varsigma)\frac{I_1\left[-q\sqrt{v^2t^2-(x-\varsigma)^2}\right]}{\sqrt{v^2t^2-(x-\varsigma)^2}} d\varsigma \\ &+ \frac{1}{2v}\int_0^t \int_{x-v(t-t')}^{x+v(t-t')} G(\varsigma,t') I_0\left[-q\sqrt{v^2(t-t')^2-(x-\varsigma)^2}\right] dt' d\varsigma. \end{aligned} \tag{3.59}$$

When $q = 0$ equation (3.53) is the forced thermal equation

$$\frac{1}{v^2}\frac{\partial^2 u}{\partial t^2} - \frac{\partial^2 u}{\partial x^2} = G(x,t). \tag{3.60}$$

On the other hand one can say that equation (3.60) is distortion-less hyperbolic equation. The condition $q = 0$ can be rewritten as:

$$V\tau = \frac{\hbar}{8} \tag{3.61}$$

The equation (3.61) is the analogous to the Heisenberg uncertainty relation. Considering equation (3.45) equation (3.61) can be written as:

$$V = \frac{T_h}{8}, \qquad V < T_h. \tag{3.62}$$

It can be stated that distortion-less waves can be generated only if $T_h > V$. For $T_h < V$, i.e. when the "Heisenberg rule" is broken, the shape of the thermal waves is changed.

In this chapter we developed the *relativistic transport equation* for an attosecond laser pulse interaction with matter. It is shown that the equation obtained is the *Proca* equation, well known in relativistic electrodynamics for massive photons. As the *heatons* are massive particles the analogy is well founded. Considering that for an attosecond laser pulse the damped term in Eq. (3.51) tends to 1, the transport phenomena are well described by the *Proca* equation.

Chapter 4

FIELD PULSE TRANSPORT IN NANOSCALE

Clusters and aggregates of atoms in the nanometer range (currently called nanoparticles) are systems intermediate in several respects, between simple molecules and bulk materials and have been the subject of intensive work.

In this paragraph, we investigate the relaxation phenomena in nanoparticles – microtubules within the frame of the quantum heat transport equation. M. Kozlowski investigated the thermal inertia of materials, heated with *electromagnetic pulse* faster than the characteristic relaxation time was investigated. It was shown, that in the case of the ultra-short laser pulses it was necessary to use the hyperbolic heat conduction (HHC). For microtubules the diameters are of the order of the de Broglie wave length. In that case quantum heat transport must be usedto describe the transport phenomena,

$$\tau \frac{\partial^2 T}{\partial t^2} + \frac{\partial T}{\partial t} = \frac{\hbar}{m} \nabla^2 T, \tag{4.1}$$

where T denotes the temperature of the heat carrier, and m denotes its mass and τ is the relaxation time. The relaxation time τ is defined as:

$$\tau = \frac{\hbar}{m v_h^2}, \tag{4.2}$$

where v_h is the pulse propagation rate

$$v_h = \frac{1}{\sqrt{3}} \alpha c \tag{4.3}$$

In equation (4.3) α is a coupling constant (for the electromagnetic interaction $\alpha = e^2 / \hbar c$ and c denotes the speed of light in vacuum. Both parameters τ and v_h characterizes completely the thermal energy transport on the atomic scale and can be termed *"atomic relaxation time"* and *"atomic"* heat diffusivity.

Both τ and v_h contain constants of Nature, α, c. Moreover, on an atomic scale there is no shorter time period than and smaller velocity than that build from of constants in Nature. Consequently, one can call τ and v_h the *elementary relaxation time* and *elementary diffusivity,* which characterizes heat transport in the elementary building block of matter, the atom. In the following, starting with elementary τ and v_h, we shall describe thermal relaxation processes in microtubules which consist of the N components (molecules) each with elementary τ and v_h. With this in view, we use the Pauli-Heisenberg inequality [1]

$$\Delta r \Delta p \geq N^{\frac{1}{3}}\hbar, \tag{4.4}$$

where r denotes the characteristic dimension of the nanoparticle and p is the momentum of the energy carriers. The Pauli-Heisenberg inequality expresses the basic property of the N – fermionic system. In fact, compared to the standard Heisenberg inequality

$$\Delta r \Delta p \geq \hbar, \tag{4.5}$$

we observe that, in this case the presence of the large number of identical fermions forces the system either to become spatially more extended for a fixed typical momentum dispension, or to increase its typical momentum dispension for a fixed typical spatial extension. We could also say that for a fermionic system in its ground state, the average energy per particle increases with the density of the system.

An illustrative means of interpreting the Pauli-Heisenberg inequality is to compare Eq. (4.4) with Eq. (4.5) and to think of the quantity on the right hand side of it as the *effective fermionic Planck constant*

$$\hbar^f(N) = N^{\frac{1}{3}}\hbar. \tag{4.6}$$

We could also say that antisymmetrization, which typifies fermionic amplitudes amplifies those quantum effects which are affected by the Heisenberg inequality.

Based on equation (4.6), we can recalculate the relaxation time τ, equation (4.2) and the thermal speed v_h, equation (4.3) for a nanoparticle consisting of N fermions

$$\hbar \leftarrow \hbar^f(N) = N^{\frac{1}{3}}\hbar \tag{4.7}$$

and obtain

$$v_h^f = \frac{e^2}{\hbar^f(N)} = \frac{1}{N^{\frac{1}{3}}} v_h, \tag{4.8}$$

$$\tau^f = \frac{\hbar^f}{m\left(v_h^f\right)^2} = N\tau. \tag{4.9}$$

The number N particles in a nanoparticle (sphere with radius r) can be calculated using the equation (we assume that density of a nanoparticle does not differ too much from that of the bulk material)

$$N = \frac{\frac{4\pi}{3} r^3 \rho AZ}{\mu} \tag{4.10}$$

and for non spherical shapes with semi axes a, b, c

$$N = \frac{\frac{4\pi}{3} abc\rho AZ}{\mu} \tag{4.11}$$

where ρ is the density of the nanoparticle, A is the Avogardo number, μ is the molecular mass of theparticles in grams and Z is the number of valence electrons.

Using equations (4.8) and (4.9), we can calculate the de Broglie wave length λ_B^f and mean free path λ_{mfp}^f for nanoparticles

$$\lambda_B^f = \frac{\hbar^f}{m v_{th}^f} = N^{\frac{2}{3}} \lambda_B, \tag{4.12}$$

$$\lambda_{emfp}^f = v_{th}^f \tau_{th}^f = N^{\frac{2}{3}} \lambda_{mfp}, \tag{4.13}$$

where λ_B and λ_{mfp} denote the de Broglie wave length and the mean free path for heat carriers in nanoparticles (e.g. microtubules). Microtubules are essential to cell functions. In neurons, microtubules help and regulate synaptic activity responsible for learning and cognitive function. Whereas microtubules have traditionally been considered to be purely structural elements, recent evidence has revealed that mechanical, chemical and electrical signaling and a communication function also exist as a result of the microtubule interaction with membrane structures by linking proteins, ions and voltage fields respectively. The characteristic dimensions of the microtubules; a crystalline cylinder 10 nm internal diameter, are of the order of the de Broglie length for electrons in atoms. When the characteristic length of the structure is of the order of the de Broglie wave length, then the signaling phenomena must be described by the quantum transport theory. In order to describe quantum transport phenomena in microtubules it is necessary to use equation (1) with the relaxation time described by equation

$$\tau = \frac{2\hbar}{mv^2} = \frac{\hbar}{E}. \tag{4.14}$$

The relaxation time is the de-coherence time, i.e. the time before the wave function collapses, when the transition classical $\rightarrow$quantum phenomena is considered.

In the following we consider the time τ for atomic and multiatomic phenomena.

$$\tau_a \approx 10^{-17}\,\mathrm{s} \tag{4.15}$$

and when we consider multiatomic transport phenomena, with N equal number of aggregates involved the equation is (4.9)

$$\tau_N = N\tau_a \tag{4.16}$$

The Penrose – Hameroff Orchestrated Objective Reduction Model (OrchOR) [4.2] proposes that quantum superposition – computation occurs in nanotubule automata within brain neurons and glia. Tubulin subunits within microtubules act as qubits, switching between states on a nanosecond scale, governed by London forces in hydrophobic pockets. These oscillations are tuned and orchestrated by microtubule associated proteins (MAPs) providing a feedback loop between the biological system and the quantum state. These qubits interact computationally by non-localquantum entanglement, according to the Schrödinger equation with preconscious processing continuing until the threshold for objective reduction (OR) is reached $(E = \hbar / T)$. At that instant, collapse occurs, triggering a "moment of awareness" or a conscious event – an event that determines particular configurations of Planck scale experiential geometry and corresponding classical states of nanotubules automata that regulate synaptic and other neural functions. A sequence of such events could provide a forward flow of subjective time and stream of consciousness. Quantum states in nanotubules may link to those in nanotubules in other neurons and glia by tunneling through gap functions, permitting extension of the quantum state through significant volumes of the brain.

Table 4.1. The de-coherence relaxation time

Event	T [ms]	E	N number of aggregates	T [ms]
Buddhist moment of awareness nucleons	13	$4\cdot 10^{15}$	10^{15}	10
Coherent 40 Hz oscillations	25	$2\cdot 10^{15}$	10^{15}	10
EEG alpha rhytm (8 to 12 Hz)	100	10^{14}	10^{14}	1
Libet's sensory threshold	100	10^{14}	10^{14}	1

Based on $E = \hbar / T$, the size and extension of Orch OR events which correlate with a subjective or neurophysiological description of conscious events can be calculated. In Table 4.1 the calculated T (Penrose-Hameroff) and τ – equation (4.16) are presented

We shall now develop the generalized quantum heat transport equation for microtubules which also includes the potential term. Thus, we are able to use the analogy of the

Schr odinger and quantum heat transport equations. If we consider, for the moment, the parabolic heat transport equation with the second derivative term omitted

$$\frac{\partial T}{\partial t} = \frac{\hbar}{m} \nabla^2 T. \tag{4.17}$$

If the real time $t \to it/2,\ T \to \Psi$, Eq. (4.17) has the form of a free Schrödinger equation

$$i\hbar \frac{\partial \Psi}{\partial t} = -\frac{\hbar^2}{2m} \nabla^2 \Psi. \tag{4.18}$$

The complete Schrödinger equation has the form

$$i\hbar \frac{\partial \Psi}{\partial t} = -\frac{\hbar^2}{2m} \nabla^2 \Psi + V\Psi, \tag{4.19}$$

where V denotes the potential energy. When we go back to real time $t \to 2it,\ \Psi \to T$, the new parabolic heat transport is obtained

$$\frac{\partial T}{\partial t} = \frac{\hbar}{m} \nabla^2 T - \frac{2V}{\hbar} T. \tag{4.20}$$

Equation (4.20) describes the quantum heat transport for $\Delta t > \tau$. For heat transport initiated by ultra-short laser pulses, when $\Delta t < \tau$ one obtains the generalized quantum hyperbolic heat transport equation

$$\tau \frac{\partial^2 T}{\partial t^2} + \frac{\partial T}{\partial t} = \frac{\hbar}{m} \nabla^2 T - \frac{2V}{\hbar} T. \tag{4.21}$$

Considering that $\tau = \hbar/mv^2$, Eq. (4.21) can be written as follows:

$$\frac{1}{v^2} \frac{\partial^2 T}{\partial t^2} + \frac{m}{\hbar} \frac{\partial T}{\partial t} + \frac{2Vm}{\hbar^2} T = \nabla^2 T. \tag{4.22}$$

Equation (4.22) describes the heat flow when apart from the temperature gradient, the potential energy V (is present.)

In the following, we consider one-dimensional heat transfer phenomena, i.e

$$\frac{1}{v^2} \frac{\partial^2 T}{\partial t^2} + \frac{m}{\hbar} \frac{\partial T}{\partial t} + \frac{2Vm}{\hbar^2} T = \frac{\partial^2 T}{\partial x^2}. \tag{4.23}$$

We seek a solution in the form

$$T(x,t) = e^{t/2\tau} u(x,t). \tag{4.24}$$

for the quantum heat transport equation (4.23)

After substitution of Eq. (4.24) into Eq. (4.23), one obtains

$$\frac{1}{v^2}\frac{\partial^2 u}{\partial t^2} - \frac{\partial^2 u}{\partial x^2} + qu(x,t) = 0. \tag{4.25}$$

where

$$q^2 = \frac{2Vm}{\hbar^2} - \left(\frac{mv}{2\hbar}\right)^2 \tag{4.26}$$

In the following, we consider a constant potential energy $V = V_0$. The general solution of Eq. (4.25) for the Cauchy boundary conditions,

$$u(x,0) = f(x), \qquad \left[\frac{\partial u(x,t)}{\partial t}\right]_{t=0} = F(x), \tag{4.27}$$

has the form [3]

$$u(x,t) = \frac{f(x-vt) + f(x+vt)}{2} + \frac{1}{2v}\int_{x-vt}^{x+vt} \Phi(x,y,z)dz, \tag{4.28}$$

where

$$\begin{aligned} \Phi(x,t,z) &= \frac{1}{v} J_0\left(\frac{b}{v}\sqrt{(z-x)^2 - v^2t^2}\right) + btf(z)\frac{J_0\left(\frac{b}{v}\sqrt{(z-x)^2 - v^2t^2}\right)}{\sqrt{(z-x)^2 - v^2t^2}}, \\ b &= \left(\frac{mv^2}{2\hbar}\right) - \frac{2Vm}{\hbar^2}v^2 \end{aligned} \tag{4.29}$$

and $J_0(z)$ denotes the Bessel function of the first kind. Considering equations (4.24), (4.25), (4.26) the solution of Eq. (4.23) describes the propagation of the distorted thermal quantum waves with characteristic lines $x = \pm vt$. We can define the distortionless thermal wave as the wave which preserves the shape in the potential energy V_0 field. The condition for conserving the shape can be expressed as

$$q^2 = \frac{2Vm}{\hbar^2} - \left(\frac{mv}{2\hbar}\right)^2 \tag{4.30}$$

When Eq. (4.30) holds, Eq. (4.31) has the form

$$\frac{\partial^2 u(x,t)}{\partial t^2} = v^2 \frac{\partial^2 u}{\partial x^2}. \tag{4.31}$$

Equation (4.31) is the quantum wave equation with the solution (for Cauchy boundary conditions (4.27))

$$u(x,t) = \frac{f(x-vt)+f(x+vt)}{2} + \frac{1}{2v}\int_{x-vt}^{x+vt} F(z)dz. \tag{4.32}$$

It is interesting to observe, that condition (4.30) has an analog in classical theory of the electrical transmission line. In the context of the transmission of an electromagnetic field, the condition $q = 0$ describes the Heaviside distortionless line. Eq. (4.30) – the distortionless condition – can be written as

$$V_0 \tau \approx \hbar, \tag{4.33}$$

We can conclude, that in the presence of the potential energy V_0 one can observe the undisturbed quantum thermal wave in microtubules only when *the Heisenberg uncertainty relation for thermal processes* (4.33) is fulfilled.

The generalized quantum heat transport equation (GQHT) (4.23) leads to generalized Schrödinger equation for microtubules. After the substitution $t \to it/2,\ T \to \Psi$ in Eq. (4.23), one obtains the generalized Schrödinger equation (GSE)

$$i\hbar\frac{\partial \Psi}{\partial t} = -\frac{\hbar^2}{2m}\nabla^2\Psi + V\Psi - 2\tau\hbar\frac{\partial^2 \Psi}{\partial t^2}. \tag{4.34}$$

Considering that $\tau = \hbar/mv^2 = \hbar/m\alpha^2c^2$ *(*$\alpha = 1/137$) is the fine-structure constant for electromagnetic interactions) Eq. (4.34) can be written as

$$i\hbar\frac{\partial \Psi}{\partial t} = -\frac{\hbar^2}{2m}\nabla^2\Psi + V\Psi - \frac{2\hbar^2}{m\alpha^2c^2}\frac{\partial^2 \Psi}{\partial t^2}. \tag{4.35}$$

One can conclude, that for a time period $\nabla t < \hbar/m\alpha^2c^2 \approx 10^{-17}$ s the description of quantum phenomena needs some revision. On the other hand, for $\nabla t > 10^{-17}$ in GSE the second derivative term can be omitted and as a result the Schrödinger equation SE is obtained, i.e.

$$i\hbar\frac{\partial\Psi}{\partial t} = -\frac{\hbar^2}{2m}\nabla^2\Psi + V\Psi \tag{4.36}$$

It is interesting to observe, that GSE was discussed also in the context of the sub-quantal phenomena.

In conclusion a study of the interactions of the attosecond laser pulses with matter can shed light on the applicability of the SE in a study of ultra-short sub-quantal phenomena.

The structure of Eq. (4.25) depends on the sign of the parameter q. For quantum heat transport phenomena with electrons as the heat carriers the parameter q is a function of the potential barrier height V_0 and velocity v.

The initial Cauchy condition

$$u(x,0) = f(x), \qquad \frac{\partial u(x,0)}{\partial t} = g(x), \tag{4.37}$$

and the solution of the Eq. (4.25) has the form [1]

$$\begin{aligned} u(x,t) &= \frac{f(x-vt)+f(x+vt)}{2} \\ &+ \frac{1}{2v}\int_{x-vt}^{x+vt} g(\varsigma) I_0\left[\sqrt{-q(v^2t^2-(x-\varsigma)^2)}\right]d\varsigma \\ &- \frac{\left(v\sqrt{-q}\right)t}{2}\int_{x-vt}^{x+vt} f(\varsigma)\frac{I_1\left[\sqrt{-q(v^2t^2-(x-\varsigma)^2)}\right]}{\sqrt{v^2t^2-(x-\varsigma)^2}}d\varsigma. \end{aligned} \tag{4.38}$$

When $q > 0$ Eq. (4.25) is the *Klein – Gordon equation* (K-G), which is well known from applications in elementary particle and nuclear physics.

For the initial Cauchy condition (4.37), the solution of the (K-G) equation can be written as

$$\begin{aligned} u(x,t) &= \frac{f(x-vt)+f(x+vt)}{2} \\ &+ \frac{1}{2v}\int_{x-vt}^{x+vt} g(\varsigma) J_0\left[\sqrt{q(v^2t^2-(x-\varsigma)^2)}\right]d\varsigma \\ &- \frac{\left(v\sqrt{q}\right)t}{2}\int_{x-vt}^{x+vt} f(\varsigma)\frac{J_0'\left[\sqrt{-q(v^2t^2-(x-\varsigma)^2)}\right]}{\sqrt{v^2t^2-(x-\varsigma)^2}}d\varsigma. \end{aligned} \tag{4.39}$$

Both solutions (4.38) and (4.39) exhibit the domains of dependence and influence on the *modified Klein-Gordon* and *Klein-Gordon equation.* These domains, which characterize the maximum speed at which a thermal disturbance can travel are determined by the principal terms of the given equation (i.e., the second derivative terms) and do not depend on the lower order terms. It can be concluded that these equations and the wave equation (for $m = 0$) have identical domains of dependence and influence.

Vacuum energy is a consequence of the quantum nature of the electromagnetic field, which is composed of photons. A photon of frequency ω has an energy $\hbar\omega$, where $\hbar$ is Planck constant. The quantum vacuum can be interpreted as the lowest energy state (or ground state) of the electromagnetic (EM) field which result when all charges and currents have been removed and the temperature has been reduced to absolute zero. In this state no ordinary photons are present. Nevertheless, because the electromagnetic field is a quantum system the energy of the ground state of the EM is not zero. Although the average value of the electric field $\langle E \rangle$ vanishes in ground state, the Root Mean Square of the field $\langle E^2 \rangle$ is not zero. Similarly the $\langle B^2 \rangle$ is not zero. Therefore the electromagnetic field energy $\langle E^2 \rangle + \langle B^2 \rangle$ is not equal zero. A detailed theoretical calculation tells that EM energy in each mode of oscillation with frequency ω is 0.5 $\hbar\omega$, which equals one half of amount energy that would be present if a single "real" photon of that mode were present. Adding up 0.5 $\hbar\omega$ for each possible states of the electromagnetic field result in a very large number for the vacuum energy E_0 in a quantum vacuum

$$E_0 = \sum_i \frac{1}{2}\hbar\omega_i. \tag{4.40}$$

The resulting vacuum energy E_0 is *infinity* unless a high frequency cut off is applied.

Inserting surfaces into the vacuum causes the states of the EM field to change. This change in the states takes place because the EM field must meet the appropriate boundary conditions at each surface. The surfaces alter the modes of oscillation and therefore alter the energy density of the lowest state of the EM field. In actual practice the change in E_0 is

$$\Delta E_0 = E_0 - E_S \tag{4.41}$$

where E_0 is the energy in empty space and E_S is the energy in space with surfaces, i.e.

$$\Delta E_0 = \frac{1}{2}\sum_n^{\substack{\text{empty}\\\text{space}}} \hbar\omega_n - \frac{1}{2}\sum_i^{\substack{\text{surface}\\\text{present}}} \hbar\omega_i\,. \tag{4.42}$$

As an example let us consider a hollow conducting rectangular cavity with sides a_1, a_2, a_3. In this case for uncharged parallel plates with an area A the attractive force between the plates is,

$$F_{att} = -\frac{\pi^2 \hbar c}{240 d^4} A, \tag{4.43}$$

where d is the distance between plates. The force F_{att} is called the parallel plate Casimir force, which was measured in three different experiments.

Recent calculations show that for conductive rectangular cavities the vacuum forces on a given face can be repulsive (positive), attractive (negative) or zero depending on the ratio of the sides.

The first measurement of repulsive Casimir force was performed by Maclay. For a distance of separation of $d \sim 0.1$ μm the repulsive force is of the order of 0.5 μN (micronewton)– for cavity geometry. In March 2001, scientist at Lucent Technology used the attractive parallel plate Casimir force to actuate a MEMS torsion device [1]. Other MEMS (MicroElectroMechanical System) have been also proposed

Standard Klein – Gordon equation is expressed as:

$$\frac{1}{c^2}\frac{\partial^2 \Psi}{\partial t^2} - \frac{\partial^2 \Psi}{\partial x^2} + \frac{m^2 c^2}{\hbar^2}\Psi = 0. \tag{4.44}$$

In equation (4.44) Ψ is the relativistic wave function for particle with mass m, c is the speed of light and $\hbar$ is Planck constant. In case of massless particles $m = 0$ and Eq. (4.44) is the Maxwell equation for photons. As was shown by Pauli and Weiskopf, the Klein – Gordon equation describes spin, – 0 bosons, because relativistic quantum mechanical equation had to allow for creation and annihilation of particles.

In the paper by J. Marciak – Kozłowska and M. Kozłowski [1] the generalized Klein – Gordon thermal equation was developed

$$\frac{1}{v^2}\frac{\partial^2 T}{\partial t^2} - \nabla^2 T + \frac{m}{\hbar}\frac{\partial T}{\partial t} + \frac{2Vm}{\hbar^2} = 0. \tag{4.45}$$

In Eq. (4.45) T denotes temperature of the medium and v is the velocity of the temperature signal in the medium. When we extract the highly oscillating part of the temperature field,

$$T = e^{-\frac{t\omega}{2}} u(x,t), \tag{4.46}$$

where $\omega = \tau^{-1}$, and τ is the relaxation time, we obtain from Eq. (4.42) (1D case)

$$\frac{1}{v^2}\frac{\partial^2 u}{\partial t^2} - \frac{\partial^2 u}{\partial x^2} + qu(x,t) = 0, \tag{4.47}$$

where

$$q = \frac{2Vm}{\hbar^2} - \left(\frac{mv}{2\hbar}\right)^2. \tag{4.48}$$

When $q > 0$ equation (4.43) is of the form of the Klein – Gordon equation in the potential field $V(x, t)$. For $q < 0$ Eq. (4.47) is the modified Klein – Gordon equation.

Considering the existence of the attosecond laser with Δt = 1 as = 10^{-18}s, Eq. (4.47) describes the heat signaling for thermal energy transport induced by ultra-short laser pulses. In the subsequent we will consider the heat transport when V is the Casimir potential. As was shown in paper [1] the Casimir force, formula (4.43), can be repulsive sign (V) = +1 and attractive sign (V) = –1. For attractive Casimir force, $V < 0$, $q < 0$ (formula (4.44)) and equation (4.43) is the modified K-G equation. For repulsive Casimir force $V > 0$ and q can be positive or negative.

As was shown by J. Maclay for different shapes of cavities, the vacuum Casimir force can change sign. Below we consider the propagation of a thermal wave within parallel plates. For Cauchy initial condition:

$$u(x,0) = 0, \qquad \frac{\partial u(x,0)}{\partial t} = g(x) \tag{4.49}$$

the solution of Eq. (4.47) has the form

$$u(x,t) = \frac{1}{2v}\int_{x-vt}^{x+vt} g(\zeta) J_0\left(\sqrt{q\left(v^2t^2 - (x-\zeta)^2\right)}\right) d\zeta \qquad \text{for} \quad q > 0 \tag{4.50}$$

and

$$u(x,t) = \frac{1}{2v}\int_{x-vt}^{x+vt} g(\zeta) I_0\left(\sqrt{-q\left(v^2t^2 - (x-\zeta)^2\right)}\right) d\zeta \qquad \text{for} \quad q < 0 \tag{4.51}$$

REFERENCE

[1] M Kozłowski, J Marciak-Kozłowska, Attoscience, arXiv.0806.0165

Chapter 5

THE ONE-DIMENSIONAL BOLTZMANN TYPE EQUATION FOR FIELD PULSE PROPAGATION

Recently it has been shown that after optical excitation by femtosecond *electromagnetic pulse* establishment of an electron temperature by e-e scattering takes place on a few hundred femtosecond time scale in both bulk and nanostructured noble materials [1]. In noble metal clusters the electron thermalization time (relaxation time) is of the order of 200 fs. This relaxation time is much larger than the duration of the now available femtosecond optical pulses offering the unique possibility of analyzing the properties of a thermal quasi-free electron gas. Uusing a two color femtosecond pump-probe laser technique the ultrafast energy exchanges of a nonequilibrium electron gas. When the duration of the laser pulse, 25 fs, in paper is shorter than the relaxation time the parabolic Fourier equation cannot be used [1]. Instead, the new hyperbolic quantum heat transport equation is the valid equation [1]. The quantum heat transport equation is the wave damped equation for heat phenomena on the femtosecond scale.

Wave is an organized propagating imbalance. Some phenomena seem to be clearly diffusive, with no wave-like implications, heat for instance. That was consistent with experiments at the late century, but not any longer. As far back as the 1960s ballistic (wave-like) heat pulses were observed at low temperatures The idea was that heat is just the manifestation of microscopic motion. Computing the classical resonant frequencies of atoms or molecules in a lattice gives numbers of the order of 10^{13} Hz, that is in the infrared, so when molecules jiggle they give off heat. These lattice vibrations are called phonons. Phonons have both wave-like and particle-like aspects. Lattice vibrations are responsible for the transport of heat, and we know that is a diffusive phenomenon, described by the Fourier equation. However, if the lattice is cooled to near absolute zero, the mean free scattering of the phonons becomes comparable to the macroscopic size of the sample. When this happens, lattice vibrations no longer behave diffusively but are actually wave-like or thermal wave. By controlling the temperature of a sample, one can control the extent to which heat is ballistic (thermal wave) or diffusive. In essence if a heat pulse is launched into sample (by the laser pulse interaction) and if the phonons can get across the sample without scattering, they will propagates as thermal waves.

The extent to which the motion of quasiparticles (phonons) or particles is ballistic, is described by the value of the relaxation time, τ. For ballistic (wave-like) motion, $\tau \rightarrow \infty$.

The equation which is generalization of the Fourier equation (in which $\tau \rightarrow \infty$) is the Heaviside equation [6] for thermal processes:

$$\tau \frac{\partial^2 T}{\partial t^2} + \frac{\partial T}{\partial t} = D\nabla^2 T \qquad (5.1)$$

For very short relaxation time, $\tau \rightarrow 0$ we obtain from equation (1) the Fourier equation

$$\frac{\partial T}{\partial t} = D\nabla^2 T \qquad (5.2)$$

and for $\tau \rightarrow \infty$ we obtain from formula (1), the ballistic ≡ thermal wave motion:

$$\frac{1}{v^2}\frac{\partial^2 T}{\partial t^2} = \nabla^2 T. \qquad (5.3)$$

In paper [1] the quantum generalization of the Heaviside equation was obtained and solved:

$$\frac{1}{v^2}\frac{\partial^2 T}{\partial t^2} + \frac{m}{\hbar}\frac{\partial T}{\partial t} = \frac{\partial^2 T}{\partial t^2}, \qquad (5.3)$$

where $v = \alpha c$ and is the fine structure constant, c is the vacuum light velocity. In formula (4) m is the *heaton* mass [1]. *Heaton* energy is equal

$$E_h = m\alpha^2 c^2. \qquad (5.4)$$

In paper [1] Heaviside equation was obtained for the fermionic gases (electrons, nucleons, quarks). In this paper the Heaviside equation will be obtained for particles with mass m, where m is the mass of the fermion or boson. Moreover beside the elastic scattering of the particles, the creation and absorption of the heat carriers will be discussed. The new form of the discrete Heaviside equation will obtained as the result of the discretization of the one-dimensional Boltzmann equation. The solution of the discrete Boltzmann equation will be obtained for Cauchy boundary conditions, initialed by ultra-short laser pulses, i.e. for $\Delta t \leq \tau$, the relaxation time.

THE MODEL

Let us consider the one-dimensional rod (strand) which can transport "particles" – heat carriers. These particles, however may move only to the right or to the left on the rod. Moving particles may collide with the fixed scatter centra, barriers, dislocations) the probabilities of such collisions and their expected results being specified. All particles will be of the same kind, with the same energy and other physical specifications distinguishable only by their direction.

Let us define:

$u(z,t)$ = expected density of particles at z and at time t moving to the right,
$v(z,t)$ = expected density of particles at z and at time t moving to the left.

Furthermore, let

$\delta(z)$ = probability of collision occurring between a fixed scattering centrum and a particle moving between z and $z+\Delta$.

Suppose that a collision might result in the disappearance of the moving particle without new particle appearing. Such a phenomenon is called *absorption*. Or the moving particle may be reversed in direction or back-scattered. We shall agreeing that in each collision at z an expected total of $F(z)$ particles arises moving in the direction of the original particle, $B(z)$ arise going in the opposite direction.

The expected total number of right-moving particles in $z_1 \leq z \leq z_2$ at time t is

$$\int_{z_1}^{z_2} u(z,t)dz, \tag{5.5}$$

while the total number of particles passing z to the right in the time interval $t_1 \leq t \leq t_2$ is

$$w\int_{t_1}^{t_2} u(z,t)dt, \tag{5.6}$$

where w is the particles speed.

Consider the particle moving to the right and passing $z+\Delta$ in the time interval $t_1 + \Delta/w \leq t \leq t_2 + \Delta/w$:

$$w\int_{t_1+\Delta/w}^{t_2+\Delta/w} u(z+\Delta,t')dt' = w\int_{t_1}^{t_2} u\left(z+\Delta,t'+\frac{\Delta}{w}\right)dt'. \tag{5.7}$$

These can arise from particles which passed z in the time interval $t_1 \leq t \leq t_2$ and came through ($z, z+\Delta$) without collision

$$w\int_{t_1}^{t_2} (1-\Delta\delta(z,t'))u(z,t')dt' \tag{5.8}$$

plus contributions from collisions in the interval $(z, z+\Delta)$. The right-moving particles interacting in $(z, z+\Delta)$ produce in the time t_1 to t_2,

$$w\int_{t_1}^{t_2} \Delta\delta(z,t')F(z,t')u(z,t')dt' \tag{5.9}$$

particles to the right, while the left moving ones give:

$$w\int_{t_1}^{t_2} \Delta\delta(z,t')B(z,t')v(z,t')dt'. \tag{5.10}$$

Thus

$$\begin{aligned} w\int_{t_1}^{t_2} u\left(z+\Delta, t'+\frac{\Delta}{w}\right)dt' = w\int_{t_1}^{t_2} u(z,t')dt' + w\Delta\int_{t_1}^{t_2} \delta(z,t')(F(z,t')-1)u(z,t')dt' \\ +w\Delta\int_{t_1}^{t_2} \delta(z,t')B(z,t')v(z,t')dt'. \end{aligned} \tag{5.11}$$

Now, we can write:

$$u\left(z+\Delta, t'+\frac{\Delta}{w}\right) = u(z,t') + \left(\frac{\partial u}{\partial z}(z,t') + \frac{1}{w}\frac{\partial u}{\partial t}(z,t')\right)\Delta \tag{5.12}$$

to get

$$\int_{t_1}^{t_2} \left(\frac{\partial u}{\partial z}(z,t')\right) + \frac{1}{w}\frac{\partial u}{\partial t}(z,t')dt' = \int_{t_1}^{t_2} \delta(z,t')((F(z,t')-1)u(z,t') + B(z,t')v(z,t'))dt'. \tag{5.13}$$

On letting $\Delta \to 0$ and differentiating with respect to t_2 we find

$$\frac{\partial u}{\partial z} + \frac{1}{w}\frac{\partial u}{\partial t} = \delta(z,t)(F(z,t)-1)u(z,t) + \delta(z,t)B(z,t)v(z,t). \tag{5.14}$$

In a like manner

$$-\frac{\partial v}{\partial z}+\frac{1}{w}\frac{\partial v}{\partial t}=\delta(z,t)B(z,t)u(z,t)+\delta(z,t)(F(z,t)-1)v(z,t). \tag{5.15}$$

The system of partial differential equations of hyperbolic type (5.15,16) is the Boltzmann equation for one dimensional transport phenomena [1].

Let us define the total density for heat carriers, $\rho(z,t)$

$$\rho(z,t)=u(z,t)+v(z,t) \tag{5.16}$$

and density of heat current

$$j(z,t)=w(u(z,t)-v(z,t)). \tag{5.17}$$

Considering equations (5.15–18) one obtains

$$\frac{\partial \rho}{\partial z}+\frac{1}{w^2}\frac{\partial j}{\partial t}=\delta(z,t)u(z,t)(F(z,t)-B(z,t)-1)+\delta(z,t)v(z,t)(B(z,t)-F(z,t)+1). \tag{5.18}$$

Equation (5.19) can be written as

$$\frac{\partial \rho}{\partial z}+\frac{1}{w^2}\frac{\partial j}{\partial t}=\frac{\delta(z,t)(F(z,t)-B(z,t)-1)j}{w} \tag{5.19}$$

or

$$j=\frac{w}{\delta(z,t)(F(z,t)-B(z,t)-1)}\frac{\partial \rho}{\partial z}+\frac{1}{w\delta(z,t)(F(z,t)-B(z,t)-1)}\frac{\partial j}{\partial t}. \tag{5.20}$$

Denoting, D, diffusion coefficient

$$D=-\frac{w}{\delta(z,t)(F(z,t)-B(z,t)-1)}$$

and τ, relaxation time

$$\tau=\frac{1}{w\delta(z,t)(1-F(z,t)-B(z,t))} \tag{5.21}$$

equation (21) takes the form

$$j = -D\frac{\partial \rho}{\partial z} - \tau\frac{\partial j}{\partial t}. \tag{5.22}$$

Equation (5.23) is the *Cattaneo's* type equation and is the generalization of the Fourier equation. Now in a like manner we obtain from equation (15–18)

$$\frac{1}{w}\frac{\partial j}{\partial z} + \frac{1}{w}\frac{\partial \rho}{\partial t} = \delta(z,t)u(z,t)(F(z,t) - 1 + B(z,t)) + \delta(z,t)v(z,t)(B(z,t) + F(z,t) - 1)) \tag{5.23}$$

or

$$\frac{\partial j}{\partial z} + \frac{\partial \rho}{\partial t} = 0. \tag{5.24}$$

Equation (5.25) describes the conservation of energy in the transport processes.

Considering equations (5.23) and (5.25) for the constant D and τ the hyperbolic Heaviside equation is obtained:

$$\tau\frac{\partial^2 \rho}{\partial t^2} + \frac{\partial \rho}{\partial t} = D\frac{\partial^2 \rho}{\partial z^2}, \tag{5.25}$$

In the case of the *heaton* gas with temperature $T(z,t)$ equation (5.26) has the form

$$\tau\frac{\partial^2 T}{\partial t^2} + \frac{\partial T}{\partial t} = D\frac{\partial^2 T}{\partial z^2},$$

where τ is the relaxation time for the thermal processes.

The Solution of the Boltzmann Equation for the Stationary Transport Phenomena in One Dimensional Strand

In the stationary state transport phenomena $dF(z,t)/dt = dB(z,t)dt = 0$ and $d\delta(z,t)/dt = 0$. In that case we denote $F(z,t) = F(z) = B(z,t) = B(z) = k(z)$ and equation (5.10) and (5.11) can be written as

$$\frac{du}{dz} = \delta(z)(k-1)u(z) + \delta(z)kv(z),$$
$$-\frac{dv}{dz} = \delta(z)k(z)u(z) + \delta(z)(k(z)-1)v(z) \tag{5.26}$$

with diffusion coefficient

$$D = \frac{w}{\delta(z)} \tag{5.27}$$

and relaxation time

$$\tau(z) = \frac{1}{w\delta(z)(1-2k(z))}. \tag{5.28}$$

The system of equations (5.27) can be written as

$$\frac{d^2u}{dz^2} - \frac{\frac{d}{dz}(\delta k)}{\delta k}\frac{du}{dz} + u\left[\delta^2(2k-1) + \frac{d\delta}{dz}(1-k) + \frac{\delta(k-1)}{\delta k}\frac{d(\delta k)}{dz}\right] = 0, \tag{5.29}$$

$$\frac{du}{dz} = \delta(k-1)u + \delta kv(z). \tag{5.30}$$

Equation (30) after differention has the form

$$\frac{d^2u}{dz^2} + f(z)\frac{du}{dz} + g(z)u(z) = 0, \tag{5.31}$$

where

$$f(z) = -\frac{1}{\delta}\left(\frac{\delta}{k}\frac{dk}{dz} + \frac{d\delta}{dz}\right),$$
$$g(z) = \delta^2(z)(2k-1) - \frac{\delta}{k}\frac{dk}{dz}. \tag{5.32}$$

For the constant absorption rate we put

$$k(z) = k = \text{constant} \neq \frac{1}{2}.$$

In that case

$$f(z) = -\frac{1}{\delta}\frac{d\delta}{dz},$$
$$g(z) = \delta^2(z)(zk-1). \tag{5.33}$$

With functions $f(z)$ and $g(z)$ the general solution of the equation (5.30) has the form

$$u(z) = C_1 e^{(1-2k)^{1/2}\int \delta dz} + C_2 e^{-(1-2k)^{1/2}\int \delta dz}. \tag{5.34}$$

In the subsequent we will consider the solution of the equation (532) with $f(z)$ and $g(z)$ described by (5.34) for Cauchy condition:

$$u(0) = q, \quad v(a) = 0. \tag{5.35}$$

Boundary condition (5.36) describes the generation of the heat carriers (by illuminating the left end of the strand with laser pulses) with velocity q heat carrier per second.

The solution has the form:

$$u(z) = \frac{2qe^{[f(0)-f(a)]}}{1+\beta e^{2[f(0)-f(a)]}}\left[\frac{(1-2k)^{\frac{1}{2}}}{(1-2k)^{\frac{1}{2}}-(k-1)}\right]\cosh[f(x)-f(a)]$$
$$+\frac{k-1}{(1-2k)^{\frac{1}{2}}-(k-1)}\sinh[f(x)-f(a)], \tag{5.36}$$
$$u(z) = \frac{2qe^{(f(0)-f(a))}}{1+\beta e^{2[f(0)-f(a)]}}\left[\frac{(1-2k)^{\frac{1}{2}}+(k-1)}{k}\sinh[f(x)-f(a)]\right],$$

where

$$f(z) = (1-2k)^{\frac{1}{2}}\int \delta dz,$$
$$f(0) = (1-2k)^{\frac{1}{2}}\left[\int \delta dz\right]_0,$$
$$f(a) = (1-2k)^{\frac{1}{2}}\left[\int \delta dz\right]_a,$$
$$\beta = \frac{(1-2k)^{\frac{1}{2}}+(k-1)}{(1-2k)^{\frac{1}{2}}-(k-1)}.$$

Considering formulae (5.17), (5.18) and (5.37) we obtain for the density, $\rho(z)$ and current density $j(z)$.

$$j(z)=\frac{2qwe^{[f(0)-f(a)]}}{1+\beta e^{2[f(0)-f(a)]}}\left[\begin{array}{l}\frac{(1-2k)^{\frac{1}{2}}}{(1-2k)^{\frac{1}{2}}-(k-1)}\cosh\left[f(z)-f(a)\right]- \\ \frac{1-2k}{(1-2k)^{\frac{1}{2}}-(k-1)}\sinh\left[f(z)-f(a)\right]\end{array}\right] \tag{5.37}$$

and

$$q=\frac{2qe^{[f(0)-f(a)]}}{1+\beta e^{2[f(0)-f(a)]}}\left[\begin{array}{l}\frac{(1-2k)^{\frac{1}{2}}}{(1-2k)^{\frac{1}{2}}-(k-1)}\cosh\left[f(z)-f(a)\right]- \\ \frac{1}{(1-2k)^{\frac{1}{2}}-(k-1)}\sinh\left[f(z)-f(a)\right]\end{array}\right]. \tag{5.38}$$

Equations (539) and (5.40) fulfill the generalized Fourier relation

$$j=-\frac{w}{\delta(z)}\frac{\partial\rho}{\partial z}, \qquad D=\frac{W}{\delta(z)}, \tag{5.39}$$

where D denotes the diffusion coefficient.

Analogously we define the generalized diffusion velocity $v_D(z)$

$$v_D(z)=\frac{j(z)}{n(z)}=$$
$$\frac{w(1-2k)^{\frac{1}{2}}\left[\cosh\left[f(z)-f(a)\right]-(1-2k)^{\frac{1}{2}}\sinh\left[f(x)-f(a)\right]\right]}{(1-2k)^{\frac{1}{2}}\cosh\left[f(x)-f(a)\right]-\sinh\left[f(x)-f(a)\right]}. \tag{5.40}$$

Assuming constant cross section for heat carriers scattering $\delta(z)=\delta_o$ we obtain from formula (5.38)

$$\begin{aligned} f(z)&=(1-2k)^{\frac{1}{2}}z, \\ f(0)&=0, \\ f(a)&=(1-2k)^{\frac{1}{2}}a \end{aligned} \tag{5.41}$$

and for density $\rho(z)$ and current density $j(z)$

$$j(z)=\frac{2qwe^{-(1-2k)^{\frac{1}{2}}a\delta}}{1+\beta e^{-(1-2k)^{\frac{1}{2}}a\delta}}\left[\frac{(1-2k)^{\frac{1}{2}}}{(1-2k)^{\frac{1}{2}}-(k-1)}\cosh\left[(2k-1)^{\frac{1}{2}}(x-a)\delta\right]\right.$$
$$\left.-\frac{(1-2k)}{(1-2k)^{\frac{1}{2}}-(k-1)}\sinh\left[(2k-1)^{\frac{1}{2}}(x-a)\delta\right]\right], \tag{5.42}$$

$$\rho(z)=\frac{2qe^{-(1-2k)^{\frac{1}{2}}a\delta}}{1+\beta e^{-(1-2k)^{\frac{1}{2}}a\delta}}\left[\frac{(1-2k)^{\frac{1}{2}}}{(1-2k)^{\frac{1}{2}}-(k-1)}\cosh\left[(2k-1)^{\frac{1}{2}\delta}(x-a)\right]\right.$$
$$\left.-\frac{1}{(1-2k)^{\frac{1}{2}}-(k-1)}\sinh\left[(2k-1)^{\frac{1}{2}}(x-a)\delta\right]\right]. \tag{5.43}$$

We define Fourier's diffusion velocity $v_F(z)$ and diffusion length, L

$$v_F=\left(\frac{D}{\tau}\right)^{\frac{1}{2}}, \qquad L=v_F\tau. \tag{5.44}$$

Considering formulae (5.28) and (5.29) one obtains

$$v_F(z)=w(1-2k)^{\frac{1}{2}},$$
$$L=\frac{1}{\delta(1-2k)^{\frac{1}{2}}}=\frac{\lambda_{\text{mfp}}}{(1\text{-}2\text{k})^{\frac{1}{2}}},$$

where λ_{mfp} denotes the mean free path for heat carriers.

Considering formulae (5.44), (5.45), (5.46), (5.47) one obtains

$$j(z)=\frac{2qwe^{-\frac{a}{L}}}{1+\beta e^{-\frac{a}{L}}}\left[\frac{(1-2k)^{\frac{1}{2}}}{(1-2k)^{\frac{1}{2}}-(k-1)}\cosh\left[\frac{(x-a)}{L}\right]\right.$$
$$\left.-\frac{(1-2k)}{(1-2k)^{\frac{1}{2}}-(k-1)}\sinh\left[\frac{x-a}{L}\right]\right], \tag{5.45}$$

$$\rho(z) = \frac{2qe^{-\frac{a}{L}}}{1+\beta e^{-\frac{a}{L}}}\left[\frac{(1-2k)^{\frac{1}{2}}}{(1-2k)^{\frac{1}{2}}-1}\cosh\left[\frac{x-a}{L}\right] - \frac{1}{(1-2k)^{\frac{1}{2}}-(k-1)}\sinh\left[\frac{x-a}{L}\right]\right]. \quad (5.46)$$

Recently [1], the heat conduction in one-dimensional system is actively investigated. As was discussed in papers [1] the dependence of density current on L can be described by the general formula

$$j \sim L^{\alpha},$$

where α can be positive or negative. The same conclusion can be drawn from the calculation presented in our paper. In this calculation coefficient α depends on the scattering cross section for the heat carriers.

REFERENCE

[1] M Kozłowski, J Marciak-Kozłowska, *Attoscience,* arXiv.0806.01656

Chapter 6

THE TWO MODE KLEIN-GORDON EQUATION FOR FIELD PULSE PROPAGATION

Dynamics of nonequilibrium electrons and phonons in metals, semiconductors have been the focus of much attention because of their fundamental interest in solid state physics and nanotechnology.

In metals, relaxation dynamics of optically excited nonequilibrium electrons has been extensively studied by pump – probe techniques using femtosecond lasers

Recently [1] it was shown that the optically excited metals relax to equilibrium with two models: rapid electron relaxation and slow thermal relaxation through the creation of the optical phonons.

In this paper we develop the hyperbolic thermal diffusion equation with two models: electrons and phonons relaxation. These two modes are characterized by two relaxation times τ_1 for electrons and τ_2 for phonons. This new equation is the generalization of our one mode hyperbolic equation, with only electrons degrees of freedom, τ_1 [1]. The hyperbolic two mode equation is the analogous equation to Klein – Gordon equation and allows the heat propagation with finite speed.

THE MODEL

As was shown in paper [1] for high frequency laser pulses the diffusion velocity exceeds that of light. This is not possible and merely demonstrates that Fourier equation is not really correct. Oliver Heaviside was well aware of this writing [2] “All diffusion formulae (as in heat conduction) show instantaneous action to an infinite distances of a source, though only to an infinitesimal extent. To make the theory of heat diffusion be rational as well as practical some modification of the equation to remove the instantaneity, however little difference it may make quantatively, in general.

August 1876 saw the appearance in Philosophical Magazine the paper which extended the mathematical understanding the diffusion (Fourier) equation. O. Heaviside for the first time wrote the hyperbolic diffusion equation for the voltage $V(x,t)$. assuming a uniform resistance, capacitance and inductance per unit length, *k, c* and *s* respectively he arrived at:

$$\frac{\partial^2 V(x,t)}{\partial x^2} = kc\frac{\partial V(x,t)}{\partial t} + sc\frac{\partial^2 V(x,t)}{\partial t^2} \tag{6.1}$$

The discussion of the broad sense of the Heaviside equation (1) can be find out, for example in our monograph [3], viz,

$$\tau^2 \frac{\partial^2 T}{\partial t^2} + \tau \frac{\partial T}{\partial t} + \frac{2V\tau}{\hbar} T = \tau \frac{\hbar}{m} \nabla^2 T \tag{6.2}$$

In Eq. (6.2) $T(\vec{r},t)$ denotes the temperature field, V is the external potential, m is the mass of heat carrier and τ is the relaxation time

$$\tau = \frac{\hbar}{mv^2} \tag{6.3}$$

As can be seen from formulae (6.2) and (6.3) in hyperbolic diffusion equation the same relaxation time τ is assumed for both type of motion: wave and diffusion.

This can not be so obvious. For example let us consider the simpler harmonic oscillator equation:

$$m\frac{d^2 x}{dt^2} + kx + c\frac{dx}{dt} = 0 \tag{6.4}$$

Equation (6.4) can be written as

$$\tau^2 \frac{d^2 x}{dt^2} + \tau \frac{dx}{dt} + x = 0 \tag{6.5}$$

where

$$\tau^2 = \frac{m}{k}, \qquad \tau = \frac{c}{k} \tag{6.6}$$

i.e.

$$c^2 = km \tag{6.7}$$

As it was well known equation (6.5) with formula (6.6) describes only the weakly damped (periodic) motion of the harmonic oscillator. It must be stressed that for HO exists also critically damped and overdamped modes which are not describes by the equation (5).

The general master equation for HO must be written as

$$\tau_1^2 \frac{d^2x}{dt^2} + \tau_2 \frac{dx}{dt} + x = 0 \tag{6.8}$$

Following the discussion of the formulae (6.5) to (6.8) we argue that the general hyperbolic diffusion equation can be written as:

$$\tau_1^2 \frac{\partial^2 T}{\partial t^2} + \tau_2 \frac{\partial T}{\partial t} + \frac{2V\tau_2}{\hbar} T = \frac{\hbar}{m} \tau_2 \nabla^2 T \tag{6.9}$$

and $\tau_1 \neq \tau_2$

Equation (9) describes the temperature field generated by ultra-short laser pulses. In Eq. (6.9) two modes: wave and diffusion are described by different relaxation times.

For quantum hyperbolic equation (9) we seek solution in the form (in 1D)

$$T(x,t) = e^{-\frac{t\tau_2}{2\tau_1^2}} u(x,t) \tag{6.10}$$

After substitution Eq. (6.10) into Eq. (6.9) one obtains

$$\frac{1}{v^2}\frac{\partial^2 u}{\partial t^2} - \frac{\partial^2 u}{\partial x^2} + qu(x,t) = 0 \tag{6.11}$$

where

$$v^2 = \frac{\hbar \tau_2}{m\tau_1^2}, \qquad q = \left(\frac{2Vm}{\hbar^2} - \frac{1}{4}\frac{m}{\hbar}\frac{\tau_2}{\tau_1^2}\right) \tag{6.12}$$

Equation (6.11) is the thermal two-mode Klein – Gordon equation and is the generalization of Klein – Gordon one mode equation developed in our monograph.

For Cauchy initial condition

$$u(x,0) = f(x), \qquad \frac{\partial u(x,0)}{\partial t} = g(x) \tag{6.13}$$

the solution of Eq. (6.11) has the form

$$
\begin{aligned}
u(x,t) = {} & \frac{f(x-vt)+f(x+vt)}{2} \\
& + \frac{1}{2v}\int_{x-vt}^{x+vt} g(\varsigma) I_0\left[\sqrt{-q\left(v^2t^2-(x-\varsigma)^2\right)}\right] d\varsigma \\
& + \frac{\left(v\sqrt{-q}\right)t}{2}\int_{x-vt}^{x+vt} f(\varsigma)\frac{I_1\left[\sqrt{-q\left(v^2t^2-(x-\varsigma)^2\right)}\right]}{\sqrt{v^2t^2-(x-\varsigma)^2}} d\varsigma
\end{aligned}
\tag{6.14}
$$

for $q < 0$

and

$$
\begin{aligned}
u(x,t) = {} & \frac{f(x-vt)+f(x+vt)}{2} \\
& + \frac{1}{2v}\int_{x-vt}^{x+vt} g(\varsigma) J_0\left[\sqrt{q\left(v^2t^2-(x-\varsigma)^2\right)}\right] d\varsigma \\
& - \frac{\left(v\sqrt{q}\right)t}{2}\int_{x-vt}^{x+vt} f(\varsigma)\frac{J_1\left[\sqrt{q\left(v^2t^2-(x-\varsigma)^2\right)}\right]}{\sqrt{v^2t^2-(x-\varsigma)^2}} d\varsigma
\end{aligned}
\tag{6.15}
$$

for $q > 0$.

REFERENCES

[1] M Kozłowski, J Marciak-Kozłowska, Attoscience, arXiv.0806.0165.
[2] O. Heaviside, Electromagnetic Theory vol. 2.

Chapter 7

NON-LINEAR KLEIN-GORDON EQUATION FOR NANOSCALE FIELD PULSE PROPAGATION

The study of transport mechanisms at the nanoscale level is of great importance nowadays. Specifically, the nanoparticles and nanotubules have important physical applications for nano- and micro-scale technologies [1]. Many models have been developed in the simple picture of point-like particles. One possibility that has been considered in the literature is that of nonlinear Klein-Gordon system where the on-site potential is ratchet-like. The development of the ultra-short electromagnetic pulses opens new possibilities in the study of the dynamics of the electrons in nanoscale systems: carbon nanotubes, nanoparticles. For attosecond laser pulses the duration of the pulse is shorter than the relaxation time. In that case the transport equations contain the second order partial derivative in time. The master equation is the Klein-Gordon equation.

In this paragraph we consider the non - linear Klein Gordon equation for mass and thermal energy transport in nanoscale. Considering the results of the monograph [1] we develop the nonlinear Klein Gordon equation for heat and mass transport in nanoscale. For ultrashort laser pulse the nonlinear Klein-Gordon equation is reduced to the nonlinear d'Alembert equation. In this paper we find out the implicit solution of the nonlinear d'Alembert equation for heat transport on nanoscale. It will be shown that for ultra-short laser pulses the non-linear Klein-Gordon equation has the nonlinear traveling wave solution.

In monograph [1] it was shown that in the case of the ultra short laser pulses the heat transport is described by the Heaviside hyperbolic heat transport equation:

$$\tau \frac{\partial^2 T}{\partial t^2} + \frac{\partial T}{\partial t} = D \frac{\partial^2 T}{\partial x^2}, \tag{7.1}$$

where T denotes the temperature of the electron gas in nanoparticle, τ is the relaxation time, m is the electron mass and D is the thermal diffusion coefficient. The relaxation time τ is defined as:

$$\tau = \frac{\hbar}{m\upsilon^2}, \quad \upsilon = \alpha c, \tag{7.2}$$

where υ is the thermal pulse propagation speed. For electromagnetic interaction when scatters are the relativistic electrons, τ = Thomson relaxation time

$$\tau = \frac{\hbar}{mc^2}. \tag{7.3}$$

Both parameters τ and υ completely characterize the thermal energy transport on the atomic scale and can be named as "atomic" relaxation time and "atomic" heat velocity.

In the following, starting with the atomic τ and υ we describe thermal relaxation processes in nanoparticles which consist of N light scatters. To that aim we use the Pauli-Heisenberg inequality:

$$\Delta r \Delta p \geq N^{\frac{1}{3}}\hbar, \tag{7.4}$$

where r denotes the radius of the nanoparticle and p is the momentum of energy carriers.

According to formula (7.4) we recalculate the relaxation time τ for nanoparticle consisting N electrons:

$$\hbar^N(N) = N^{\frac{1}{3}}\hbar, \tag{7.5}$$

$$\tau^N = N\tau. \tag{7.6}$$

Formula (7.6) describes the scaling of the relaxation time for N *fermion* systems.

With formulae (7.4) and (7.5) the heat transport equation takes the form:

$$\tau^N \frac{\partial^2 T}{\partial t^2} + \frac{\partial T}{\partial t} = \frac{\hbar^{\frac{1}{3}}}{m}\frac{\partial^2 T}{\partial x^2} \tag{7.7}$$

and for mass transport:

$$\tau^N \frac{\partial^2 N}{\partial t^2} + \frac{\partial N}{\partial t} = \frac{N^{\frac{1}{3}}\hbar}{m}\frac{\partial^2 N}{\partial x^2}. \tag{7.8}$$

Equation (7.7) is linear damped Klein-Gordon equation, and was solved for nanotechnology systems in [1].

The nonlinearity of Eq. (7.8) opens new possibilities for the study of non-stationary stable processes in molecular nanostructures. Let us consider equation (7.8) in more details:

$$(N\tau)\frac{\partial^2 N}{\partial t^2} + \frac{\partial N}{\partial t} = \frac{N^{\frac{1}{3}}\hbar}{m}\frac{\partial^2 N}{\partial x^2}. \tag{7.9}$$

For $\Delta t < N\tau$ Eq (9) is the nonlinear d'Alembert equation

$$\frac{1}{\upsilon^2}\frac{\partial^2 N}{\partial t^2} = \frac{\partial^2 N}{\partial x^2} \qquad (7.10)$$

with N dependent velocity:

$$\upsilon = \frac{1}{N^{\frac{1}{3}}}\alpha c. \qquad (7.11)$$

Equation (7.10) can be written in more general form:

$$\frac{\partial^2 N}{\partial t^2} = \frac{\partial G(N)}{\partial x}, \qquad (7.12)$$

where

$$G(N) = f(N)\frac{\partial N}{\partial t}. \qquad (7.13)$$

The traveling wave solution of equation (12) has implicit form [1]:

$$\lambda^2 N(x,t) - \int G(N)dN = A(x + \lambda t) + B, \qquad (7.14)$$

where A,B and λ are arbitrary constants.

REFERENCE

[1] M Kozłowski, J Marciak-Kozłowska, Attoscience, arXiv.0806.0165

Chapter 8

SUB- AND SUPERSONIC FIELD PULSE MOTION

Recently it has become possible to produce MeV electrons with short-pulse multiteravat electromagnetic field pulse [1]. The fast ignitor concept relevant to the inertial confinement fusion enhances the interest in this process. In an under-dense plasma, electrons and ions tend to be expelled from the focal spot by the ponderomotive pressure of an intense electromagnetic pulse, and the formed channel can act as a propagation guide for the laser beam. Depending on the quality of the laser beam, the cumulative effects of ponderomotive and relativistic self focusing can significantly increase the laser intensity. For these laser pulses, the laser electric and magnetic fields reach few hundreds of GV/m and megagauss, respectively, and quiver velocity in the laser field is closed to the light speed. The component of the resulting Lorentz force $\left(e\vec{v}\times\vec{B}\right)$ accelerates electrons in the longitudinal direction, and energies of several tens of MeV can be achieved Recently the spectra of hot electrons (i.e. with energy in MeV region) were investigated. The interaction of 500 fs FWHM pulses with CH target produces electrons with energy up to 20 MeV were observed. Moreover for electrons with energies higher than 5 MeV the change of electron temperature was observed: from 1 MeV (for energy of electrons < 5 MeV) to 3 MeV (for energy of electrons > 5 MeV). In this paper the interaction of femtosecond laser pulse with electron plasma will be investigated. Within the theoretical framework of Heaviside temperature wave equation, the heating process of the plasma will be described. It will be shown that in vicinity of energy of 5 MeV the sound velocity in plasma reaches the value $\frac{c}{\sqrt{3}}$ and is independent of the electron energy.

THE MODEL

The mathematical form of the hyperbolic quantum heat transport was proposed in [1] Under the absence of heat or mass sources the equations can be written as the Heaviside equations:

$$\frac{1}{v_\rho^2}\frac{\partial^2\rho}{\partial t^2}+\frac{1}{D_\rho}\frac{\partial\rho}{\partial t}=\frac{\partial^2\rho}{\partial x^2} \tag{8.1}$$

and

$$\frac{1}{\upsilon_T^2}\frac{\partial^2 T}{\partial t^2}+\frac{1}{D_T}\frac{\partial T}{\partial t}=\frac{\partial^2 T}{\partial x^2} \tag{8.2}$$

for mass and thermal energy transport respectively. The discussion of the properties of Eq. (8.1,8.2) was performed in [1] In Eq. (8.1) υ_ρ is the velocity of density wave, D_ρ is the diffusion coefficient for mass transfer. In Eq. (8.2) υ_T is the velocity for thermal energy propagation and D_T is the thermal diffusion coefficient.

In the subsequent we will discuss the complex transport phenomena, i.e. diffusion and convection in the external field. The current density in the case when the diffusion and convection are taken into account can be written as:

$$j=-D_\rho\frac{\partial\rho}{\partial t}-\tau_\rho\frac{\partial j}{\partial t}+\rho V. \tag{8.3}$$

In equation (8.3) the first term describes the Fourier diffusion, the second term is the Maxwell – Cattaneo term and the third term describes the convection with velocity V. The continuity equation for the transport phenomena has the form:

$$\frac{\partial j}{\partial x}+\frac{\partial\rho}{\partial t}=0. \tag{8.4}$$

Considering both equations (8.3) and (8.4) one obtains the transport equation:

$$\frac{\partial\rho}{\partial t}=-\tau_\rho\frac{\partial^2\rho}{\partial t^2}+D_\rho\frac{\partial^2\rho}{\partial x^2}-V\frac{\partial\rho}{\partial x}. \tag{8.5}$$

In equation (8.5) τ_ρ denotes the relaxation time for transport phenomena. Let us perform the Smoluchowski transformation for $\rho(x,t)$

$$\rho=\exp\left[\frac{Vx}{2D}-\frac{V^2t}{4D}\right]\rho_1(x,t). \tag{8.6}$$

After substituting $\rho(x,t)$ formula (6) to equation (8.5) one obtains for $\rho_1(x,t)$:

$$\tau_\rho\frac{\partial^2\rho_1}{\partial t^2}+\left(1-\tau_\rho\frac{V_\rho^2}{2D_\rho}\right)\frac{\partial\rho_1}{\partial t}+\tau_\rho\frac{V_\rho^4}{16D_\rho^2}\rho_1=D_\rho\frac{\partial^2\rho_1}{\partial x^2}. \tag{8.7}$$

Considering that $D_\rho=\tau_\rho\upsilon_\rho^2$ equation (8.7) can be written as

$$\tau_\rho \frac{\partial^2 \rho_1}{\partial t^2} + \left(1 - \frac{V_\rho^2}{2\upsilon_\rho^2}\right)\frac{\partial \rho_1}{\partial t} + \frac{1}{16\tau_\rho}\frac{V_\rho^4}{\upsilon_\rho^4}\rho_1 = D_\rho \frac{\partial^2 \rho_1}{\partial x^2}. \tag{8.8}$$

In the same manner equation for the temperature field can be obtained:

$$\tau_T \frac{\partial^2 T_1}{\partial t^2} + \left(1 - \frac{V_T^2}{2\upsilon_T^2}\right)\frac{\partial T_1}{\partial t} + \frac{1}{16\tau_T}\frac{V_T^4}{\upsilon_T^4}T_1 = D_T \frac{\partial^2 T_1}{\partial x^2}. \tag{8.9}$$

In equation (9) τ_T, D_T, V_T and υ_T are: relaxation time for heat transfer, diffusion coefficient, heat convection velocity and thermal wave velocity.

Subsequently we will investigate the structure and solution of the equation (8.9). For the hyperbolic heat transport Eq. (8.9) we seek a solution of the form:

$$T_1(x,t) = e^{-\frac{t}{2\tau_T}} u(x,t). \tag{8.10}$$

After substitution of Eq. (8.10) into Eq. (8.9) one obtains:

$$\begin{aligned} &\tau_T \frac{\partial^2 u(x,t)}{\partial t^2} - D_T \frac{\partial^2 u(x,t)}{\partial x^2} + \left(-\frac{1}{4\tau_T} + \frac{V_T^2}{4D_T} + \tau_T \frac{V_T^4}{16D_T^2}\right)u(x,t) \\ &-\tau_T \frac{V_T^2}{2D_T}\frac{\partial u(x,t)}{\partial t} = 0. \end{aligned} \tag{8.11}$$

Considering that $D_T = \tau_T \upsilon_T^2$ Eq. (8.11) can be written as

$$\begin{aligned} &\tau_T \frac{\partial^2 u}{\partial t^2} - \tau_T \upsilon_T^2 \frac{\partial^2 u}{\partial x^2} + \left(-\frac{1}{4\tau_T} + \frac{V_T^2}{4\tau_T \upsilon_T^2} + \tau_T \frac{V_T^4}{16\tau_T^2 \upsilon_T^4}\right)u(x,t) \\ &-\tau_T \frac{V_T^2}{2\upsilon_T^2}\frac{\partial u}{\partial t} = 0. \end{aligned} \tag{8.12}$$

After omitting the term $\frac{V_T^4}{\upsilon_T^4}$ in comparison to the term $\frac{V_T^2}{\upsilon_T^2}$ Eq. (8.12) takes the form:

$$\frac{\partial^2 u}{\partial t^2} - \upsilon_T^2 \frac{\partial^2 u}{\partial x^2} + \frac{1}{4\tau_T^2}\left(-1 + \frac{V_T^2}{\upsilon_T^2}\right)u(x,t) - \frac{V_T^2}{2\upsilon_T^2 \tau_T}\frac{\partial u}{\partial t} = 0. \tag{8.13}$$

Considering that $\tau_T^{-2} >> \tau_T^{-1}$ one obtains from Eq. (8.13)

$$\frac{\partial^2 u}{\partial t^2} - \upsilon_T^2 \frac{\partial^2 u}{\partial x^2} + \frac{1}{4\tau_T^2}\left(-1 + \frac{V_T^2}{\upsilon_T^2}\right) u(x,t) = 0. \tag{8.14}$$

Equation (14) is the master equation for heat transfer induced by ultra-short laser pulses, i.e. when $\Delta t \approx \tau_T$. In the following we will consider the Eq. (8.14) in the form:

$$\frac{\partial^2 u}{\partial t^2} - \upsilon_T^2 \frac{\partial^2 u}{\partial x^2} - qu(x,t) = 0, \tag{8.15}$$

where

$$q = \frac{1}{4\tau_T^2}\left(\frac{V_T^2}{\upsilon_T^2} - 1\right). \tag{8.16}$$

In equation (816) the ratio

$$M_T = \frac{V_T}{\upsilon_T} = \frac{V_T}{\upsilon_S} \tag{8.17}$$

is the Mach number for thermal processes, for $\upsilon_T = \upsilon_S$ is the sound velocity in the gas of heat carriers. In monograph [1] the structure of equation (8.15) was investigated. It was shown that for $q < 0$, i.e. $V_T < \upsilon_S$, subsonic heat transfer is described by the modified telegrapher equation

$$\frac{1}{\upsilon_T^2}\frac{\partial^2 u}{\partial t^2} - \frac{\partial^2 u}{\partial x^2} + \frac{1}{4\tau_T^2\upsilon_T^2}\left(\frac{V_T^2}{\upsilon_S^2} - 1\right) u(x,t) = 0. \tag{8.18}$$

For $q > 0$, $V_T > \upsilon_S$, i.e. for supersonic case heat transport is described by Klein-Gordon equation:

$$\frac{1}{\upsilon_T^2}\frac{\partial^2 u}{\partial t^2} - \frac{\partial^2 u}{\partial x^2} + \frac{1}{4\tau_T^2\upsilon_T^2}\left(\frac{V_T^2}{\upsilon_S^2} - 1\right) u(x,t) = 0. \tag{8.19}$$

The velocity of sound υ_S depends on the temperature of the heat carriers. The general formula for sound velocity reads [1]:

$$\upsilon_S^2 = \left(zG - \frac{G}{z}\left(1 + \frac{5G}{z} - G^5\right)^{-1}\right)^{-1}. \tag{8.20}$$

In formula (8.20) $z = \frac{mc^2}{T}$ and G is of the form [1]:

$$G = \frac{K_3(z)}{K_2(z)}, \tag{8.21}$$

where c is the light velocity, m is the mass of heat carrier, T is the temperature of the gas and $K_3(z)$, $K_2(z)$ are modified Bessel functions of the second kind.

For the initial conditions

$$u(x,0) = f(x), \qquad \left[\frac{\partial u(x,t)}{\partial t}\right]_{t=0} = F(x). \tag{8.22}$$

Solution of the equation can be find in [1]:

$$u(x,t) = \frac{f(x - \upsilon_T t) + f(x + \upsilon_T t)}{2} + \frac{1}{2}\int_{x-\upsilon_T t}^{x+\upsilon_T t} \Phi(x,t,z)dz, \tag{8.23}$$

where

$$\Phi(x,t,z) = \frac{1}{\upsilon_T} F(z) J_0\left(\frac{\sqrt{q}}{\upsilon_T}\sqrt{(z-x)^2 - \upsilon_T^2 t^2}\right) +$$

$$\sqrt{q}tf(z)\frac{J_0'\left(\frac{\sqrt{q}}{\upsilon_T}\sqrt{(z-x)^2 - \upsilon_T^2 t^2}\right)}{\sqrt{(z-x)^2 - \upsilon_T^2 t^2}}$$

and

$$\sqrt{q} = \frac{1}{4\tau_T^2}\left(\frac{V_T^2}{\upsilon_T^2} - 1\right).$$

The general equation for complex heat transfer: diffusion plus convection can be written as:

$$\frac{\partial T}{\partial t} = -\tau_T \frac{\partial^2 T}{\partial t^2} + D_T \frac{\partial^2 T}{\partial x^2} - V_T \frac{\partial T}{\partial x}. \tag{8.24}$$

Considering Eqs. (8.6), (8.10) and (8.23) the solution of equation (24) is

$$T(x,t)=\exp\left[\frac{V_T x}{2D}-\frac{V_T^2 t}{4D}\right]\exp\left[-\frac{t}{2\tau_T}\right]\cdot\left(\begin{array}{l}\dfrac{f(x-\upsilon_T t)+f(x+\upsilon_T t)}{2}+\\ \dfrac{1}{2}\displaystyle\int_{x-\upsilon_T t}^{x+\upsilon_T t}\Phi(x,t,z)dz\end{array}\right).$$

REFERENCE

[1] M Kozłowski, J Marciak-Kozłowska, *Attoscience*, arXiv.0806.0165

Chapter 9

THE FIELD WAVES IN N-DIMENSIONAL SPACE

The fact that we perceive the world to have three spatial dimensions is some-thing so familiar to our experience of its structure that we seldom pause to consider the direct influence this special property has upon the laws of physics. Yet some have done so and there have been many intriguing at-tempts to deduce the expediency or inevitability of a three-dimensional world from the general structure of the physical law themselves.

As earlier as 1917 P. Ehrenfest] pointed out that neither classical atoms nor planetary orbits can be stable in a space with $n > 3$ and traditional quantum atoms cannot be stable either As far as $n < 3$ is concerned, it has been argued that organism would face insurmountable topological problem if $n = 2$: for instance two nerves cannot across. In the following we will conjecture that since $n = 2$ offers vastly less complexity that $n = 3$, worlds with $n < 3$ are just too simple and barren to contain observers. Since our Universe appears governed by the propagation of classical and quantum waves it is interesting elucidate the nature of the connection the properties of the wave equation and the spatial dimensions.

In this paragraph we describe the partial differential equation (PDE) for the propagation of the thermal waves in n-dimensional space time. It is well known that for heat transport induced by ultra-short laser pulses (shorter than the relaxation time) the governing equation can be written as [1]

$$\frac{1}{\upsilon^2}\frac{\partial^2 T}{\partial t^2} + \frac{1}{D}\frac{\partial T}{\partial t} + \frac{2Vm}{\hbar^2}T = \nabla^2 T, \tag{9.1}$$

where T is the temperature, υ denotes the thermal wave propagation, m is the mass of heat carriers and V is the potential.

In paper [1] the solution of the equation for one-dimensional case, $n = 1$ was obtained. In this paper we develop and solve the analog of the equation for n = natural numbers $n = 1$, 2,..., separately for n = odd and n = even. The Huygens' principle for thermal wave will be discussed. It will be shown that for thermal waves only in odd dimensional space the waves propagate at exactly a fixed space velocity υ without "echoes" assuming the absence of walls (potentials) or inhomogeneities.

The three-dimensional heat transfer induced by ultra-short laser pulses in Cu_3Au alloy will be suggested.

THE MASTER EQUATION FOR THE THERMAL WAVES IN N-DIMENSIONS

In the following we consider the n-dimensional heat transfer phenomena described by the equation [1]

$$\frac{1}{\upsilon^2}\frac{\partial^2 T}{\partial t^2}+\frac{1}{D}\frac{\partial T}{\partial t}+\frac{2Vm}{\hbar^2}T=\nabla^2 T, \tag{9.2}$$

where temperature T is the function in the n-dimensional space

$$T=T(x_1,x_2,...,x_n,t). \tag{9.3}$$

We seek solution of equation (2) in the form:

$$T(x_1,x_2,...,x_n,t)=e^{-\frac{t}{2\tau}}u(x_1,x_2,...,x_n,t). \tag{9.4}$$

After substitution of Eq. (4) to Eq. (2) one obtains

$$\frac{1}{\upsilon^2}\frac{\partial^2 u}{\partial t^2}-\nabla^2 u+qu=0, \tag{9.5}$$

where

$$q=\frac{2Vm}{\hbar^2}-\left(\frac{m\upsilon}{2\hbar}\right)^2$$

for $D=\frac{\hbar}{m}$

We can define the distortionless thermal wave as the wave which preserves the shape in the field of the potential V. The condition for conserving the shape can be formulated as

$$q=\frac{2Vm}{\hbar^2}-\left(\frac{m\upsilon}{2\hbar}\right)^2=0. \tag{9.6}$$

When Eq. (6) holds Eq. (5) has the form

$$\frac{1}{\upsilon^2}\frac{\partial^2 u}{\partial t^2}-\nabla^2 u=0 \tag{9.7}$$

and condition (6) can be written as

$$V\tau \sim \hbar. \tag{9.8}$$

We conclude that in the presence of the potential energy V one can observe the undisturbed thermal wave only when the Heisenberg uncertainty relation (8) is fulfilled.

The solution of the Eq. (7) for the n-odd can be find in [1]. First of all let us change the variables in Eq. (7)

$$\upsilon : t \to t', \quad x \to x', \quad u \to u'$$

and obtain

$$\frac{\partial^2 u'}{\partial t'^2} - \nabla^2 u' = 0. \tag{9.9}$$

For

$$\begin{aligned} &\lim_{(x',t')\to(x^0,0)} u'(x',t') = g(x_0), \\ &\lim_{(x',t')\to(x^0,0)} \frac{\partial u'(x',t')}{\partial t} = h(x_0), \end{aligned} \tag{9.10}$$

the solution has the form [5]

$$u'(x',t') = \frac{1}{\gamma_n}\left[\begin{array}{l} \left(\frac{\partial}{\partial t'}\right)\left(\frac{1}{t'}\frac{\partial}{\partial t'}\right)^{\frac{n-3}{2}}\left(t'^{n-2}\oint\limits_{\partial B(x',t')} g\,dS\right) \\ +\left(\frac{1}{t'}\frac{\partial}{\partial t'}\right)^{\frac{n-3}{2}}\left(t'^{n-2}\oint\limits_{\partial B(x',t')} h\,dS\right)\end{array}\right] \tag{9.11}$$

and $\gamma_n = 1\cdot 3\cdot 5\cdots(n-2)$.

For n – even the solution of equation (9.1) has the form [1].

$$u'(x',t') = \frac{1}{\gamma_n}\left[\begin{array}{l} \left(\frac{\partial}{\partial t'}\right)\left(\frac{1}{t'}\frac{\partial}{\partial t'}\right)^{\frac{n-2}{2}}\left(t'^{n}\oint\limits_{\partial B(x',t')} \frac{g(y')dy'}{\left(t'^2-|y'-x'|^2\right)^{\frac{1}{2}}}\right) \\ +\left(\frac{1}{t'}\frac{\partial}{\partial t'}\right)^{\frac{n-2}{2}}\left(t'^{n}\oint\limits_{\partial B(x',t')} \frac{h(y')dy'}{\left(t'^2-|y'-x'|^2\right)^{\frac{1}{2}}}\right)\end{array}\right] \tag{9.12}$$

and $\gamma_n = 2 \cdot 4 \cdots (n-2) \cdot n$. In formulae (9.11) and (19.12) $\oint$ denotes integral over n – space. Considering formulae (9.11) and (9.12) we conclude that for n-odd the solution (9.11) is dependent on the value of functions h and g only on the hypersphere $\partial B(x',t')$. On the other hand for n-even the solution (9.12) is dependent on the values of the functions h and g on the full hyperball $B(x',t')$. In the other words for n-odd $n \geq 3$ the value of the initial functions h and g influences the solution (912) only on the surface of the cone $\{(y',t'),t'>0,|x'-y'|=t'\}$. For n = even the value of the functions g and h influences the solution on the full cone. It means that the thermal wave induced by the disturbance for n = odd have the well defined front. For n-even the wave influences space after the transmission of the front. This means that Huygens' principle is false for n-even. In conclusion: if we solve the wave equation in n-dimensions the signals propagate sharply (i.e. Huygens' principle is valid) only for dimensions n = 3, 5, 7, ... Thus three is the "best of all possible" dimensions, the smallest dimension in which signals propagate sharply.

SUGGESTIONS FOR EXPERIMENTALISTS

The hyperbolic transport equation for heat transport (9.13)

$$\frac{1}{\upsilon_T^2}\frac{\partial^2 T}{\partial t^2} + \frac{1}{D_T}\frac{\partial T}{\partial t} + \frac{2Vm}{\hbar^2}T = \nabla^2 T \tag{9.13}$$

or mass transport

$$\frac{1}{\upsilon_\rho^2}\frac{\partial^2 \rho}{\partial t^2} + \frac{1}{D_\rho}\frac{\partial \rho}{\partial t} + \frac{2Vm}{\hbar^2}\rho = \nabla^2 \rho \tag{9.14}$$

are the damped wave equations. For very short time period $\Delta t \sim \tau$ equations (9.13) and (9.14) can be written as

$$\frac{1}{\upsilon_{\rho,T}^2}\frac{\partial^2 u_{\rho,T}}{\partial t^2} - \nabla^2 u_{\rho,T} = 0. \tag{9.15}$$

Eq. (9.15) is the generalization of equation (9.7). The solution of equation (15) in n-dimensional cases are described by formulae (9.11) and (9.12).

As it is well known only in 3-dimensional case the Huygens principle is fulfilled. It seems that in order to observe the thermal wave not disturbed by the "echoes" and with sharp front *the true three-dimensional experiment* must be performed. Moreover the experiment must be performed in the relaxation regime, i.e. for materials with relatively long relaxation time. The best candidate for "relaxation materials" is the Cu_3Au alloy [6]. As was shown in [1] the relaxation time is of the order of 10^4 s in the temperature range 650 - 660 K. For $t >$

660 K the abrupt increasing, up to $1.5 \cdot 10^{5}$ s (due to order → disorder transition) was observed.

REFERENCE

[1] M Kozłowski, J Marciak-Kozłowska, *Attoscience,* arXiv.0806.0165.

Chapter 10

ULTRA-SHORT FIELD PULSE PROPAGATION IN CARBON NANOTUBES

The interaction of the laser pulses with carbon nanotubes is a very interesting and new field of investigation. In nanotechnology the carbon nanotubes are the main parts of MEMS and in the future NEMS devices. In living organisms nanotubes build the skeleton of the living cells. The exceptional properties of carbon nanotubes (CNTs), including ballistic transport and semiconducting behaviour with band-gaps in the range of 1 eV, have sparked a large number of theoretical and experimental] studies. The possibility of using CNTs to replace crystalline silicon for high-performance transistors has resulted in an effort to reduce the size of CNT field-effect transistors (CNTFETs) in order to understand the scaling behaviour and the ultimate limit. In this context we discuss CNT transistors with channel lengths less than 20 nm but which have characteristics comparable to those of much larger silicon-based field effect transistors with similar channel lengths

Since the first CNTFET was demonstrated in 1998, their characteristics have been continuously and rapidly. The breakthrough progress has been made recently in generation and detection of ultrashort, attosecond laser pulses with high harmonic generation technique. This is the beginning of the attophysics age in which many-electron dynamics will be investigated in real time. In the existing new laser projects the generation of 100 GW-level attosecond lectromagnetic X-ray pulses is investigated. With ultra-short (attosecond) high energetic laser pulses the relativistic multi-electrons states can be generated. For relativistic one electron state the Dirac equation is the master equation. In this paper we develop and solve the Dirac type thermal equation for multi-electron states generated in laser interaction with matter. In this paper the Dirac one dimensional thermal equation will be applied to study of the generation of the positron-electron pairs. It will be shown that the cross section is equal to the Thomson cross-section for the electron-electron scattering.

DERIVATION OF THE 1+1 DIMENSIONAL DIRAC EQUATION FOR THERMAL PROCESSES IN MICROTUBULES

As pointed in paper [1] spin-flip occurs only when there is more than one dimension in space. Repeating the discussion of deriving the Dirac equation for the case of one spatial dimension, one easily finds that the Dirac matrices $\boldsymbol{\alpha}$ and $\boldsymbol{\beta}$ are reduced to 2x2 matrices that can be represented by the Pauli matrices. This fact simply implies that if there is only one spatial dimension, there is no spin. It should be instructive to show explicitly how to derive the 1+1 dimensional Dirac equation. As discussed in textbooks a wave equation that satisfies relativistic covariance in space-time as well as the probabilistic interpretation should have the form:

$$i\hbar\frac{\partial}{\partial t}\Psi(x,t)=\left[c\boldsymbol{\alpha}\left(-i\hbar\frac{\partial}{\partial x}\right)+\boldsymbol{\beta}m_0c^2\right]\Psi(x,t). \tag{10.1}$$

To obtain the relativistic energy-momentum relation $E^2=(pc)^2+m_0^2c^4$ we postulate that (10.1) coincides with the Klein-Gordon equation

$$\left[\frac{\partial^2}{\partial(ct)^2}-\frac{\partial^2}{\partial x^2}+\left(\frac{m_0c}{\hbar}\right)^2\right]\Psi(x,t)=0. \tag{10.2}$$

By comparing (10.1) and (10.2) it is easily seen that $\boldsymbol{\alpha}$ and $\boldsymbol{\beta}$ must satisfy

$$\boldsymbol{\alpha}^2-\boldsymbol{\beta}^2=1, \qquad \boldsymbol{\alpha\beta}+\boldsymbol{\beta\alpha}=0. \tag{10.3}$$

Any two of the Pauli matrices can satisfy these relations. Therefore, we may choose $\boldsymbol{\alpha}=\boldsymbol{\sigma}_x$ and $\boldsymbol{\beta}=\boldsymbol{\sigma}_z$ and we obtain:

$$i\hbar\frac{\partial}{\partial t}\Psi(x,t)=\left[c\boldsymbol{\sigma}_x\left(-i\hbar\frac{\partial}{\partial x}\right)+\boldsymbol{\sigma}_z m_0c^2\right]\Psi(x,t), \tag{10.4}$$

where $\Psi(x,t)$ is a 2-component spinor.

The Eq. (10.4) is the Weyl representation of the Dirac equation. We perform a phase transformation on $\Psi(x,t)$ letting $u(x,t)=\exp\left(\frac{imc^2t}{\hbar}\right)\Psi(x,t)$. Call u's upper (respectively, lower) component $u_+(x,t)$, $u_-(x,t)$; it follows from (10.4) that $u_\pm$ satisfies

$$\frac{\partial u_\pm(x,t)}{\partial t}=\pm c\frac{\partial u_\pm}{\partial x}+\frac{im_0c^2}{\hbar}(u_\pm-u_\mp). \tag{10.5}$$

Following the physical interpretation of the equation (10.5) it describes the relativistic particle (mass m_0) propagates at the speed of light c and with a certain *chirality* (like a two component neutrino) except that at random times it flips both direction of propagation (by 180^0) and chirality.

In [1] we considered a particle moving on the line with fixed speed w and supposed that from time to time it suffers a complete reversal of direction, $u(x,t) \Leftrightarrow v(x,t)$, where $u(x,t)$ denotes the expected density of particles at x and at time t moving to the right, and $v(x,t) \equiv$ expected density of particles at x and at time t moving to the left. In the following we perform the change of the abbreviation

$$\begin{aligned} u(x,t) &\rightarrow u_+, \\ v(x,t) &\rightarrow u_-. \end{aligned} \tag{10.6}$$

Following the results of the paper [1] we obtain for the $u_\pm(x,t)$ the following equations

$$\begin{aligned} \frac{\partial u_+}{\partial t} &= -w\frac{\partial u_+}{\partial x} - \frac{w}{\lambda}((1-k)u_+ - ku_-), \\ \frac{\partial u_-}{\partial t} &= w\frac{\partial u_-}{\partial x} + \frac{w}{\lambda}(ku_+ + (k-1)u_-). \end{aligned} \tag{10.7}$$

In equation (10.7) $k(x)$ denotes the number of the particles which are moving in left (right) direction after the scattering at x. The mean free path for scattering is equal λ, $\lambda = w\tau$, where τ is the relaxation time for scattering.

Comparing formulae (10.5) and (10.7) we conclude that the shapes of both equations are the same. In the subsequent we will call the set of the equations (10.7) *the Dirac equation* for the particles with velocity w, mean free path λ. For thermal processes we define $T_{+,-} \equiv$ the temperature of the particles with chiralities + and – respectively and with analogy to equation (10.7) we obtain:

$$\begin{aligned} \frac{\partial T_+}{\partial t} &= -w\frac{\partial T_+}{\partial x} - \frac{w}{\lambda}((1-k)T_+ - kT_-), \\ \frac{\partial T_-}{\partial t} &= w\frac{\partial T_-}{\partial x} + \frac{w}{\lambda}(kT_+ + (k-1)T_-), \end{aligned} \tag{10.8}$$

where $\dfrac{w}{\lambda} = \dfrac{1}{\tau}$.

In one dimensional case we introduce one dimensional cross section for scattering

$$\sigma(x,t) = \frac{1}{\lambda(x,t)}. \tag{10.9}$$

THE SOLUTION OF THE DIRAC EQUATION FOR STATIONARY TEMPERATURES IN MICROTUBULES

In the stationary state thermal transport phenomena $\frac{\partial T_{+,-}}{\partial t}=0$ and Eq. (10.8) can be written as

$$\begin{aligned}\frac{dT_+}{dx}&=-\sigma\left((1-k)T_+ + kT_-\right),\\ \frac{dT_-}{dx}&=\sigma(k-1)T_- + \sigma kT_+.\end{aligned} \tag{10.10}$$

After the differentiation of the equation (10.9) we obtain for $T_+(x)$

$$\frac{d^2T_+}{dx}-\frac{1}{\sigma k}\frac{d}{dx}(\sigma k)\frac{dT_+}{dx}+T_+\left[\begin{matrix}\sigma^2(2k-1)+\frac{d\sigma}{dx}(1-k)+\\ \frac{\sigma(k-1)}{\sigma k}\frac{d(\sigma k)}{dx}\end{matrix}\right]=0. \tag{10.11}$$

Equation (10.10)10.11) can be written in a compact form

$$\frac{d^2T_+}{dx^2}+f(x)\frac{dT_+}{dx}+g(x)T_+=0,$$

where

$$\begin{aligned}f(x)&=-\frac{1}{\sigma}\left(\frac{\sigma}{k}\frac{dk}{dx}+\frac{d\sigma}{dx}\right),\\ g(x)&=\sigma^2(x)(2k-1)-\frac{\sigma}{k}\frac{dk}{dx}.\end{aligned} \tag{10.12}$$

In the case for constant $\frac{dk}{dx}=0$ we obtain

$$\begin{aligned}f(x)&=-\frac{1}{\sigma}\frac{d\sigma}{dx},\\ g(x)&=\sigma^2(x)(2k-1).\end{aligned}$$

With functions $f(x)$, $g(x)$ described by formula (10.12) the general solution of Eq.

$$T_+(x) = C_1 e^{(1-2k)^{\frac{1}{2}}\int \sigma(x)dx} + C_2 e^{-(1-2k)^{\frac{1}{2}}\int \sigma(x)dx} \tag{10.13}$$

and

$$T_-(x) = \frac{\left[(1-k)+(1-2k)^{\frac{1}{2}}\right]}{k} \times \left[C_1 e^{(1-2k)^{\frac{1}{2}}\int \sigma(x)dx} + \frac{(1-k)-(1-2k)^{\frac{1}{2}}}{(1-k)+(1-2k)^{\frac{1}{2}}} C_2 e^{-(1-2k)^{\frac{1}{2}}\int \sigma(x)dx} \right]. \tag{10.14}$$

The formulae (10.13) and (10.14) describe three different mode for heat transport. For $k = \frac{1}{2}$ we obtain $T_+(x) = T_-(x)$ while for $k > \frac{1}{2}$, i.e. for heat carrier generation $T_+(x)$ and $T_-(x)$ are the thermal waves. For $k < \frac{1}{2}$ i.e. for strong absorption $T_+(x)$ and $T_-(x)$ represents the diffusion motion. The mechanism of the scattering is for the moment unknown. The farreaching hypothesis is that the electrons interact with the point field (ZPF) and as the result the additional carriers are generated

In the subsequent we will consider the solution of Eq. (10.9) for Cauchy conditions:

$$T_+(0) = T_0, \quad T_-(a) = 0. \tag{10.15}$$

Boundary conditions (10.15) describes the generation of heat carriers by illuminating the left end of one dimensional slab (with length a) by laser pulse.

From formulae (10.13) and (10.14) we obtain:

$$T_+(x) = \frac{2T_0 e^{[f(0)-f(a)]}}{1+\beta e^{2[f(0)-f(a)]}} \times \frac{(1-2k)^{\frac{1}{2}} \cosh\left[f(x)-f(a)\right] + (k-1)\sinh\left[f(x)-f(a)\right]}{(1-2k)^{\frac{1}{2}} - (k-1)}, \tag{10.16}$$

$$T_-(x) = \frac{2T_0 e^{2[f(0)-f(a)]}\left[(k-1)+(1-2k)^{\frac{1}{2}}\right]\sinh[f(x)-f(a)]}{\left(1+\beta e^{-2[f(a)-f(0)]}\right)k}. \tag{10.17}$$

In formulae (10.16) and (10.17)

$$\beta = \frac{(1-2k)^{\frac{1}{2}} + (k-1)}{(1-2k)^{\frac{1}{2}} - (k-1)} \tag{10.18}$$

and

$$\begin{aligned} f(x) &= (1-2k)^{\frac{1}{2}} \int \sigma(x) dx, \\ f(0) &= (1-2k)^{\frac{1}{2}} \left[\int \sigma(x) dx \right]_0, \\ f(a) &= (1-2k)^{\frac{1}{2}} \left[\int \sigma(x) dx \right]_a. \end{aligned} \tag{10.19}$$

With formulae (10.16) for $T_+(x)$ and $T_-(x)$ we define the asymmetry $A(x)$ of the temperature $T(x)$

$$A(x) = \frac{T_+(x) - T_-(x)}{T_+(x) + T_-(x)}, \tag{10.20}$$

$$A(x) = \frac{\dfrac{(1-2k)^{\frac{1}{2}}}{(1-2k)^{\frac{1}{2}} - (k-1)} \cosh\left[f(x) - f(a)\right] - \dfrac{1-2k}{(1-2k)^{\frac{1}{2}} - (k-1)} \sinh\left[f(x) - f(a)\right]}{\dfrac{(1-2k)^{\frac{1}{2}}}{(1-2k)^{\frac{1}{2}} - (k-1)} \cosh\left[f(x) - f(a)\right] - \dfrac{1}{(1-2k)^{\frac{1}{2}} - (k-1)} \sinh\left[f(x) - f(a)\right]} \tag{10.21}$$

From formula (10.21) we conclude that for elastic scattering, i.e. when $k = \frac{1}{2}$, $A(x) = 0$, and for $k \neq \frac{1}{2}$, $A(x) \neq 0$.

In the monograph [1] we introduced the relaxation time τ for quantum heat transport

$$\tau = \frac{\hbar}{mv^2}. \tag{10.22}$$

In formula (10.22) m denotes the mass of heat carriers electrons and $v = \alpha c$, where α is the fine structure constant for electromagnetic interactions. As was shown in monograph τ is also the lifetime for positron-electron pairs in vacuum.

When the duration time of the laser pulse is shorter than τ, then to describe the transport phenomena we must use the hyperbolic transport equation. Recently the structure of water was investigated with the attosecond $(10^{-18}\,\mathrm{s})$ resolution Considering that $\tau \approx 10^{-17}$ s we argue that to study performed in [1] open the new field for investigation of laser pulse with matter. In order to apply the equations (10.9) to attosecond laser induced phenomena we must know the cross section $\sigma(x)$. Considering formula (10.22) we obtain

$$\sigma(x)=\frac{mv}{\hbar}=\frac{me^2}{\hbar^2} \tag{10.23}$$

and it occurs $\sigma(x)$ is the Thomson cross section for electron-electron scattering.

With formula (10.23) the solution of Cauchy problem has the

$$T_+(x)=$$

$$\frac{2T_0 e^{-(1-2k)^{\frac{1}{2}}\frac{me^2}{\hbar^2}a}}{\left[1+\beta e^{-2(1-2k)^{\frac{1}{2}}\frac{me^2}{\hbar^2}a}\right]}\times$$

$$\frac{(1-2k)^{\frac{1}{2}}\cosh\left[(1-2k)^{\frac{1}{2}}\frac{me^2}{\hbar^2}(x-a)\right]+(k-1)\sinh\left[(1-2k)^{\frac{1}{2}}\frac{me^2}{\hbar^2}(x-a)\right]}{(1-2k)^{\frac{1}{2}}-(k-1)},$$

$$T_-(x)=\frac{2T_0 e^{-\frac{(1-2k)^{\frac{1}{2}}me^2a}{\hbar^2}}\left[(k-1)-(1-2k)^{\frac{1}{2}}\right]}{\left(1+\beta e^{-2(1-2k)^{\frac{1}{2}}\frac{me^2}{\hbar^2}a}\right)k}\times$$

$$\sinh\left[(1-2k)^{\frac{1}{2}}\frac{me^2}{\hbar^2}(x-a)\right]. \tag{10.24}$$

REFERENCE

[1] M Kozłowski, J Marciak-Kozłowska, *Attoscience,* arXiv.0806.0165

Chapter 11

SCHRODINGER-NEWTON WAVE MECHANICS

When M. Planck made the first quantum discovery he noted an interesting fact [1]. The speed of light, Newton's gravity constant and Planck's constant clearly reflect fundamental properties of the world. From them it is possible to derive the characteristic mass M_P, length L_P and time T_P with approximate values

$L_P = 10^{-35}$ m
$T_P = 10^{-43}$ s
$M_P = 10^{-5}$ g.

Nowadays much of cosmology is concerned with "interface" of gravity and quantum mechanics.

In this paragraph we investigate the very simple question: how gravity can modify the quantum mechanics, i.e. the nonrelativistic Schrödinger equation (SE). We argue that SE with relaxation term describes properly the quantum behaviour of particle with mass $m < M_P$ and contains the part which can be interpreted as the pilot wave equation. For $m \to M_P$ the solution of the SE represent the strings with mass M_P.

GENERALIZED FOURIER LAW

The thermal history of the system (heated gas container, semiconductor or Universe) can be described by the generalized Fourier equation

$$q(t) = -\int_{-\infty}^{t} \underbrace{K(t-t')}_{\text{thermal history}} \underbrace{\nabla T(t')dt'}_{\text{diffusion}}. \tag{11.1}$$

In Eq. (11.1) $q(t)$ is the density of the energy flux, T is the temperature of the system and $K(t-t')$ is the thermal memory of the system

$$K(t-t') = \frac{K}{\tau}\exp\left[-\frac{(t-t')}{\tau}\right], \tag{11.2}$$

where K is constant, and τ denotes the relaxation time.

With formula (11.2) the hyperbolic thermal equation can be obtained

$$\frac{\partial^2 T}{\partial t^2} + \frac{1}{\tau}\frac{\partial T}{\partial t} = \frac{D_T}{\tau}\nabla^2 T. \tag{11.3}$$

For $\tau \to 0$, Eq. (11.3) is the Fourier thermal equation

$$\frac{\partial T}{\partial t} = D_T \nabla^2 T \tag{11.4}$$

and D_T is the thermal diffusion coefficient. The systems with very short relaxation time have very short memory. On the other hand for $\tau \to \infty$ Eq. (11.3) has the form of the thermal wave (undamped) equation, or *ballistic* thermal equation. In the solid state physics the *ballistic* phonons or electrons are those for which $\tau \to \infty$. The experiments with *ballistic* phonons or electrons demonstrate the existence of the *wave motion* on the lattice scale or on the electron gas scale.

$$\frac{\partial^2 T}{\partial t^2} = \frac{D_T}{\tau}\nabla^2 T. \tag{11.5}$$

For the systems with very long memory Eq. (11.3) is time symmetric equation with no arrow of time, for the Eq. (11.5)does not change the shape when $t \to -t$.

In Eq. (11.3) we define:

$$\upsilon = \left(\frac{D_T}{\tau}\right), \tag{11.6}$$

velocity of thermal wave propagation and

$$\lambda = \upsilon\tau, \tag{11.7}$$

where λ is the mean free path of the heat carriers. With formula (11.6) equation (11.3) can be written as

$$\frac{1}{\upsilon^2}\frac{\partial^2 T}{\partial t^2} + \frac{1}{\tau\upsilon^2}\frac{\partial T}{\partial t} = \nabla^2 T. \tag{11.8}$$

DAMPED WAVE EQUATION, THERMAL CARRIERS IN POTENTIAL WELL, V.

From the mathematical point of view equation:

$$\frac{1}{v^2}\frac{\partial^2 T}{\partial t^2}+\frac{1}{D}\frac{\partial T}{\partial t}=\nabla^2 T$$

is the hyperbolic partial differential equation (PDE). On the other hand Fourier equation

$$\frac{1}{D}\frac{\partial T}{\partial t}=\nabla^2 T \tag{11.9}$$

and Schrödinger equation

$$i\hbar\frac{\partial \Psi}{\partial t}=-\frac{\hbar^2}{2m}\nabla^2\Psi \tag{11.10}$$

are the parabolic equations. Formally with substitutions

$$t \leftrightarrow it,\ \Psi \leftrightarrow T \tag{11.11}$$

Fourier equation (11.9) can be written as

$$i\hbar\frac{\partial \Psi}{\partial t}=-D\hbar\nabla^2\Psi \tag{11.12}$$

and by comparison with Schrödinger equation one obtains

$$D_T\hbar=\frac{\hbar^2}{2m} \tag{11.13}$$

and

$$D_T=\frac{\hbar}{2m}. \tag{11.14}$$

Considering that $D_T=\tau v^2$ (11.6) we obtain from (11.14)

$$\tau = \frac{\hbar}{2mv_h^2}. \tag{11.15}$$

Formula (11.15) describes the relaxation time for quantum thermal processes. Starting with Schrödinger equation for particle with mass m in potential V:

$$i\hbar\frac{\partial\Psi}{\partial t} = -\frac{\hbar^2}{2m}\nabla^2\Psi + V\Psi \tag{11.16}$$

and performing the substitution (11.11) one obtains

$$\hbar\frac{\partial T}{\partial t} = \frac{\hbar^2}{2m}\nabla^2 T - VT \tag{11.17}$$

$$\frac{\partial T}{\partial t} = \frac{\hbar}{2m}\nabla^2 T - \frac{V}{\hbar}T. \tag{11.18}$$

Equation (11.18) is Fourier equation (parabolic PDE) for $\tau = 0$. For $\tau \neq 0$ we obtain

$$\tau\frac{\partial^2 T}{\partial t^2} + \frac{\partial T}{\partial t} + \frac{V}{\hbar}T = \frac{\hbar}{2m}\nabla^2 T, \tag{11.19}$$

$$\tau = \frac{\hbar}{2mv^2} \tag{11.20}$$

or

$$\frac{1}{v^2}\frac{\partial^2 T}{\partial t^2} + \frac{2m}{\hbar}\frac{\partial T}{\partial t} + \frac{2Vm}{\hbar^2}T = \nabla^2 T.$$

Model Schrödinger Equation

With the substitution (11.11) equation (11.19) can be written as

$$i\hbar\frac{\partial\Psi}{\partial t} = V\Psi - \frac{\hbar^2}{2m}\nabla^2\Psi - \tau\hbar\frac{\partial^2\Psi}{\partial t^2}. \tag{11.21}$$

The new term, relaxation term

$$\tau\hbar\frac{\partial^2\Psi}{\partial t^2} \tag{11.22}$$

describes the interaction of the particle with mass m with space-time. The relaxation time τ can be calculated as:

$$\tau^{-1} = \left(\tau_{e-p}^{-1} + ... + \tau_{Planck}^{-1}\right), \tag{11.23}$$

where, for example $\tau_{e\text{-}p}$ denotes the scattering of the particle m on the electron-positron pair ($\tau_{e-p} \sim 10^{-17}$ s) and the shortest relaxation time τ_{Planck} is the Planck time ($\tau_{Planck} \sim 10^{-43}$ s).

From equation (11.23) we conclude that $\tau \approx \tau_{Planck}$ and equation (11.21) can be written as

$$i\hbar\frac{\partial\Psi}{\partial t} = V\Psi - \frac{\hbar^2}{2m}\nabla^2\Psi - \tau_{Planck}\hbar\frac{\partial^2\Psi}{\partial t^2}, \tag{11.24}$$

where

$$\tau_{Planck} = \frac{1}{2}\left(\frac{\hbar G}{c^5}\right)^{\frac{1}{2}} = \frac{\hbar}{2M_p c^2}. \tag{11.25}$$

In formula (11.25) M_p is the mass Planck. Considering Eq. (11.25), Eq. (11.24) can be written as

$$i\hbar\frac{\partial\Psi}{\partial t} = -\frac{\hbar^2}{2m}\nabla^2\Psi + V\Psi - \frac{\hbar^2}{2M_p}\nabla^2\Psi + \frac{\hbar^2}{2M_p}\nabla^2\Psi - \frac{\hbar^2}{2M_p c^2}\frac{\partial^2\Psi}{\partial t^2}. \tag{11.26}$$

The last two terms in Eq. (11.6) can be defined as the *Bohmian* pilot wave

$$\frac{\hbar^2}{2M_p}\nabla^2\Psi - \frac{\hbar^2}{2M_p c^2}\frac{\partial^2\Psi}{\partial t^2} = 0, \tag{11.27}$$

i.e.

$$\nabla^2\Psi - \frac{1}{c^2}\frac{\partial^2\Psi}{\partial t^2} = 0. \tag{11.28}$$

It is interesting to observe that pilot wave Ψ does not depend on the mass of the particle. With postulate (11.28) we obtain from equation (11.26)

$$i\hbar\frac{\partial\Psi}{\partial t}=-\frac{\hbar^2}{2m}\nabla^2\Psi+V\Psi-\frac{\hbar^2}{2M_p}\nabla^2\Psi \tag{11.29}$$

and simultaneously

$$\frac{\hbar^2}{2M_p}\nabla^2\Psi-\frac{\hbar^2}{2M_pc^2}\frac{\partial^2\Psi}{\partial t^2}=0. \tag{11.30}$$

In the operator form Eqs. (11.9) and (11.20) can be written as

$$\hat{E}=\frac{\hat{p}^2}{2m}+\frac{1}{2M_pc^2}\hat{E}^2, \tag{11.31}$$

where $\hat{E}$ and $\hat{p}$ denote the operators for energy and momentum of the particle with mass m. Equation (11.31) is the new dispersion relation for quantum particle with mass m. From Eq. (21) one can concludes that Schrödinger quantum mechanics is valid for particles with mass $m \ll M_P$. But pilot wave exists independent of the mass of the particles.

For particles with mass $m \ll M_P$ Eq. (11.9) has the form

$$i\hbar\frac{\partial\Psi}{\partial t}=-\frac{\hbar^2}{2m}\nabla^2\Psi+V\Psi. \tag{11.32}$$

SCHRÖDINGER EQUATION AND THE STRINGS

In the case when $m \approx M_p$ Eq. (11.29) can be written as

$$i\hbar\frac{\partial\Psi}{\partial t}=-\frac{\hbar^2}{2M_p}\nabla^2\Psi+V\Psi, \tag{11.33}$$

but considering Eq. (11.30) one obtains

$$i\hbar\frac{\partial\Psi}{\partial t}=-\frac{\hbar^2}{2M_pc^2}\frac{\partial^2\Psi}{\partial t^2}+V\Psi \tag{11.34}$$

or

$$\frac{\hbar^2}{2M_pc^2}\frac{\partial^2\Psi}{\partial t^2}+i\hbar\frac{\partial\Psi}{\partial t}-V\Psi=0. \tag{11.35}$$

We look for the solution of Eq. (11.35) in the form

$$\Psi(x,t)=e^{i\omega t}u(x). \tag{11.36}$$

After substitution formula (11.16) to Eq. (11.35) we obtain

$$\frac{\hbar^2}{2M_pc^2}\omega^2+\omega\hbar+V(x)=0 \tag{11.37}$$

with the solution

$$\begin{aligned}\omega_1&=\frac{-M_pc^2+M_pc^2\sqrt{1-\dfrac{2V}{M_pc^2}}}{\hbar}\\ \omega_2&=\frac{-M_pc^2-M_pc^2\sqrt{1-\dfrac{2V}{M_pc^2}}}{\hbar}\end{aligned} \tag{11.38}$$

for $\dfrac{M_pc^2}{2}>V$ and

$$\begin{aligned}\omega_1&=\frac{-M_pc^2+iM_pc^2\sqrt{\dfrac{2V}{M_pc^2}-1}}{\hbar}\\ \omega_2&=\frac{-M_pc^2-iM_pc^2\sqrt{\dfrac{2V}{M_pc^2}-1}}{\hbar}\end{aligned} \tag{11.39}$$

for $\dfrac{M_pc^2}{2}<V$.

Both formulae (11.38) and (11.39) describe the string oscillation, formula (11.27) damped oscillation and formula (11.28) over damped string oscillation.

D. Bohm presented the pilot wave theory in 1952 and de Broglie had presented a similar theory in the mid 1920's. It was rejected in 1950's and the rejection had nothing to do with de Broglie and Bohm later works.

There is always the possibility that the pilot wave has a primitive, mind like property. That's how Bohm described it. We can say that all the particles in the Universe end even Universe have their own pilot waves, their own information. Then the consciousness for example is the very complicated receiver of the surrounding pilot wave fields.

In this paragraph a study of the Newton-Schr ̈odinger-Bohm (NSB) equation for the pilot wave was developed:

$$i\hbar\frac{\partial\Psi}{\partial t}=-\frac{\hbar^2}{2m}\nabla^2\Psi+V\Psi-\frac{\hbar^2}{2M_p}\nabla^2\Psi+\frac{\hbar^2}{2M_p}\left(\nabla^2\Psi-\frac{1}{c^2}\frac{\partial^2\Psi}{\partial t^2}\right). \tag{11.40}$$

In Eq. (1) m is the mass of the quantum particle and M_P is the Planck mass ($M_P \approx 10^{-5}$ g).

For elementary particles with mass $m << M_P$ we obtain from Eq. (1)

$$i\hbar\frac{\partial\Psi}{\partial t}=-\frac{\hbar^2}{2m}\nabla^2\Psi+V\Psi+\frac{\hbar^2}{2M_p}\left(\nabla^2\Psi-\frac{1}{c^2}\frac{\partial^2\Psi}{\partial t^2}\right) \tag{11.41}$$

and for macroscopic particles with $m >> M_P$ equation (1) has the form:

$$i\hbar\frac{\partial\Psi}{\partial t}=-\frac{\hbar^2}{2M_p}\nabla^2\Psi+\frac{\hbar^2}{2M_p}\left(\nabla^2\Psi-\frac{1}{c^2}\frac{\partial^2\Psi}{\partial t^2}\right)+V\Psi \tag{11.42}$$

or

$$i\hbar\frac{\partial\Psi}{\partial t}=-\frac{\hbar^2}{2M_pc^2}\frac{\partial^2\Psi}{\partial t^2}+V\Psi$$

and is dependent of m.

Discussion of the results. In the following we will discuss the pilot wave time evolution for the macroscopic particles, i.e. for particles with $m >> M_P$.

For V = const. we seek the solution of Eq. (11.42) in the form:

$$\Psi=e^{\gamma t}. \tag{11.43}$$

After substitution formula (11.43) to Eq. (11.42) one's obtains

$$M_P\gamma^2+\frac{2M_P^2c^2}{\hbar}\gamma-\frac{2M_P^2c^2}{\hbar^2}V=0 \tag{11.44}$$

with the solution

$$\gamma_{1,2}=-\frac{iM_Pc^2}{\hbar}\pm\frac{M_Pc^2}{\hbar}\sqrt{-1+\frac{2V}{M_Pc^2}}. \tag{11.45}$$

For a free particle, $V = 0$ we obtain:

$$\gamma_{1,2} = \begin{cases} 0, \\ -\dfrac{2M_P c^2}{\hbar} i. \end{cases} \tag{11.46}$$

According to formulae (11.43) and (11.46) equation (11.42) has the solution

$$\Psi(t) = A + Be^{-\frac{2M_P c^2 i}{\hbar} t}. \tag{11.47}$$

For $t = 0$ we put $\Psi(0) = 0$, then

$$\Psi(t) = A\left(1 - e^{-\frac{2it}{\tau_P}}\right),$$

where τ_P = Planck time

$$\tau_P = \frac{\hbar}{M_p c^2} \tag{11.48}$$

$$E = \hbar\omega = 10^{19}\ \text{GeV} \tag{11.49}$$

and period T

$$T = 10^{-42}\ \text{s}.$$

Considering that the contemporary limiting time resolution is of the order 10^{-18} s (attosecond laser pulses) the jittering of the free macroscopic particle can not be observed.

Classically, when the inertial mass m_i and the gravitational mass m_g are equated the mass drops out of Newton's equation of motion, implying that particles of different mass with the same initial condition follows the same trajectories. But in Schrödinger's equation the masses do not cancel. For example in a uniform gravitational field

$$i\hbar\frac{\partial\Psi}{\partial t} = -\frac{\hbar^2}{2m_i}\frac{\partial^2\Psi}{\partial x^2} + m_g g x\Psi$$

implying mass dependent difference in motion.

In this paragraph we investigate the motion of particle with inertial mass m_i in the potential field V. As was shown in monograph [2] the general QM equation has the form

$$i\hbar\frac{\partial\Psi}{\partial t}=V\Psi-\frac{\hbar^2}{2m}\nabla^2\Psi-2\tau\hbar\frac{\partial^2\Psi}{\partial t^2}$$

where the term

$$2\tau\hbar\frac{\partial^2\Psi}{\partial t^2},\quad \tau=\frac{\hbar}{m_i c^2}$$

describes the memory of the particle with mass m_i. Above equation for the wave function Ψ is the local equation with finite invariant speed, c which equals the light speed in the vacuum.

THE MODEL EQUATION

In the subsequent local Schrödinger equation with finite Planck mass is obtained:

$$i\hbar\frac{\partial\Psi}{\partial t}=V\Psi-\frac{\hbar^2}{2m}\nabla^2\Psi-\tau\hbar\frac{\partial^2\Psi}{\partial t^2} \tag{11.50}$$

The new relaxation term (memory term)

$$\tau\hbar\frac{\partial^2\Psi}{\partial t^2} \tag{11.51}$$

describes the interaction of the particle with mass m with space-time.

The relaxation time τ can be calculated as

$$\tau^{-1}=\left(\tau_{e-p}^{-1}+\ldots+\tau_{Planck}^{-1}\right) \tag{11.52}$$

where $\tau_{e-p}\sim 10^{-17}$ s denotes the scattering of the particle m on the electron – positron virtual pair, $\tau_{Planck}\approx 10^{-43}$ s

$$\tau_{Planck}=\frac{\hbar}{M_p c^2} \tag{11.53}$$

where M_p is Planck mass.

Eq. (11.49) can be written as:

$$i\hbar\frac{\partial\Psi}{\partial t}=-\frac{\hbar^2}{2m}\nabla^2\Psi+V\Psi-\frac{\hbar^2}{2M_p}\nabla^2\Psi+\frac{\hbar^2}{2M_p}\left(\nabla^2\Psi-\frac{1}{c^2}\frac{\partial^2\Psi}{\partial t^2}\right) \tag{11.54}$$

As can be seen from Eq (11.53) for $M_p \to \infty$ one obtains non-local Schrödinger equation

$$i\hbar\frac{\partial}{\partial t}\Psi=-\frac{\hbar^2}{2m}\nabla^2\Psi+V\Psi \tag{11.55}$$

From equation (11.53) can be concluded that Schrödinger QM is valid for particles with $m<<M_p$. The last term

$$\frac{\hbar^2}{2M_p}\left(\nabla^2\Psi-\frac{1}{c^2}\frac{\partial^2\Psi}{\partial t^2}\right) \tag{11.56}$$

when is equal zero

$$\nabla^2\Psi-\frac{1}{c^2}\frac{\partial^2\Psi}{\partial t^2}=0 \tag{11.57}$$

describes the pilot wave equation. It is interesting to observe that the pilot wave equation is independent of mass of the particles.

Let us look for the solution of the Eq. (11.53), $V=0$, in the form (for 1D)

$$\Psi=\Psi(x-ct) \tag{11.58}$$

For $\tau \neq 0$, i.e. for finite Planck mass we obtain:

$$\Psi(x-ct)=\exp\left(\frac{2\mu ic}{\hbar}(x-ct)\right) \tag{11.59}$$

where the reduced μ mass equals

$$\mu=\frac{m_i M_p}{m+M_p} \tag{11.60}$$

For $m << M_p$, i.e. for all elementary particles one obtains

$$\mu = m_i \tag{11.61}$$

and formula (11.59) describes the wave function for free Schrödinger particles

$$\Psi(x-ct)=\exp\left(\frac{2mic}{\hbar}(x-ct)\right) \tag{11.62}$$

For $m >> M_p$, $\mu = M_p$

$$\Psi(x-ct)=\exp\left(\frac{2M_p ic}{\hbar}(x-ct)\right) \tag{11.63}$$

From formula (6) we conclude that $\Psi(x-ct)$ is independent of m of particle, m. In the case $m < M_p$ from formulae (11.52,53) one obtains

$$\begin{aligned}\mu &= m\left(1-\frac{m}{M_p}\right)\\ \Psi(x-ct) &= \exp\left(\frac{2imc}{\hbar}(x-ct)\right)\exp\left(-i\frac{m}{M_p}\left(\frac{2mc}{\hbar}x-\frac{2mc^2}{\hbar}t\right)\right)\end{aligned} \tag{11.64}$$

In formula (11.64) we put

$$\begin{aligned}k &= \frac{2m_i c}{\hbar}\\ \omega &= \frac{2m_i c^2}{\hbar}\end{aligned} \tag{11.65}$$

and obtain

$$\Psi(x-ct)=e^{i(kx-\omega t)}e^{-i\frac{m}{M_p}(kx-\omega t)} \tag{11.66}$$

As can concluded from formula (11.66) the second term depends on the gravity

$$\exp\left[-i\frac{m_i}{M_p}(kx-\omega t)\right]=\exp\left[-i\left(\frac{m_i^2 G}{\hbar c}\right)^{1/2}(kx-\omega t)\right] \tag{11.67}$$

where G is the Newton gravity constant.

It is interesting to observe that the new constant, α_G,

$$\alpha_G=\frac{m_i^2 G}{\hbar c} \tag{11.68}$$

is the gravitational fine structure constant. For $m_i = m_N$ nucleon mass

$$\alpha_G = 5.9042 \cdot 10^{-39}$$

REFERENCES

[1] M. Kozlowski, J. Marciak-Kozlowska, *Attoscience,* arXiv 0806.0165
[2] J. Marciak-Kozlowska, M. Kozlowski, *From femto-to attoscience and beyond,* Nova Science Publishers, New York, USA, 2009

Chapter 12

FIELD PULSES IN QUANTUM CORRALS

Recently has been a great interest in both ultrafast (femtosecond and attosecond) laser-induced kinetics and in nanoscale properties of matter Particular attention has been attracted by phenomena that are simultaneously nanoscale and ultrafast. Fundamentally nanosize eliminates effects of electromagnetic retardation and thus facilitates coherent ultrafast kinetics. On the applied side, nanoscale design of optoelectronic devices is justified if their operating times are ultrashort to allow for ultrafast computing and transmission of information. One of the key problems of ultrafast/nanoscale physics is ultrafast excitation of nanosystem where the transferred energy localizes at a given site. Because the electromagnetic wavelength is on a much larger microscale it is impossible to employ light-wave focusing for that purpose.

In recent years the advances in scanning tunneling microscopy (STM) made possible the manipulation of single atoms on top of a surface and the construction of quantum-nanometre scale structures of arbitrary shapes. In particular, quantum corrals have been assembled by depositing a close line of atoms or molecules on Cu or noble metal surface. These surface have the property that for small wave vectors parallel to the surface a parabolic band of two-dimensional (2D) surface states uncoupled to bulk states exists. In quantum corrals the STM tip can existed standing wave pattern of the one electron de Broglie waves.

In this paper we describe the thermal excitation of the de Broglie electron waves with attosecond laser pulses. With coherent control of the ultrashort laser pulses it is possible to concentrate the laser energy on the nanometer scale. Following the results of monograph [2] we will describe the temperature of the electron 2D gas with the help of the quantum hyperbolic heat transfer equation.

THE MODEL

In the following we consider the 2D heat transfer phenomena described by the equation [2]:

$$\frac{1}{v^2}\frac{\partial^2 T}{\partial t^2}+\frac{1}{D}\frac{\partial T}{\partial t}+\frac{2Vm}{\hbar^2}T=\nabla^2 T, \tag{12.1}$$

where T is temperature of the 2D electron gas

$$T = T(x, y, t), \tag{12.2}$$

D is the thermal diffusion coefficient, V is the nonthermal potential and m is the mass of the heat carriers–electrons.

We seek solution of Eq. (10.1) in the form

$$T(x, y, t) = e^{-\frac{t}{2\tau}} u(x, y, t). \tag{12.3}$$

After substitution of Eq. (12.3) to Eq. (10.1) one obtains

$$\frac{1}{v^2}\frac{\partial^2 u}{\partial t^2} - \nabla^2 u + qu = 0, \tag{12.4}$$

where

$$q = \frac{2Vm}{\hbar^2} - \left(\frac{mv}{2\hbar}\right)^2 \tag{12.5}$$

for $D = \dfrac{\hbar}{m}$.

We can define the distortionless thermal wave as the wave which preserves the shape in the field of the potential V. The condition for conserving the shape can be formulated as

$$q = \frac{2Vm}{\hbar^2} - \left(\frac{mv}{2\hbar}\right)^2 = 0. \tag{12.6}$$

When Eq. (12.6) holds Eq. (10.4) has the form

$$\frac{1}{v^2}\frac{\partial^2 u}{\partial t^2} - \nabla^2 u = 0 \tag{12.7}$$

and condition (12.6) can be written as

$$V\tau \sim \hbar, \tag{12.8}$$

where τ is the relaxation time

$$\tau = \frac{\hbar}{mv^2}. \tag{12.9}$$

We conclude that in the presence of the potential energy V one can observe the undisturbed thermal wave only when the Heisenberg uncertainty relaxation (10.8) is fulfilled.

In the subsequent we will consider the thermal relaxation of the 2D electron gas contained in 2D circular quantum corral with the radius r. In that case in polar coordinates equation (10.7 has the form

$$\frac{1}{r}\frac{\partial}{\partial r}\left(r\frac{\partial u}{\partial r}\right) + \frac{1}{r^2}\frac{\partial^2 u}{\partial \theta^2} = \frac{1}{v^2}\frac{\partial^2 u}{\partial t^2}, \tag{12.10}$$

where $0 < r < a$, $-\pi < \theta < \pi$ with boundary condition

$$\begin{aligned} &u(r,\theta,0) = f(r,\theta), \quad 0 < r < a, \quad -\pi < \theta \le \pi, \\ &\frac{\partial u}{\partial t}(r,\theta,0) = g(r,\theta), \quad 0 < r < a, \quad -\pi < \theta \le \pi. \end{aligned} \tag{12.11}$$

Using separation of variables $u(r,t) = R(r)T(t)$ yields the solution

$$\begin{aligned} u(r,\theta,t) = &\sum_n a_{on} J_0(\lambda_{on} r)\cos(\lambda_{on} vt) \\ &+ \sum_{m,n} a_{mn} J_m(\lambda_{mn} r)\cos(m\theta)\cos(\lambda_{mn} ct) \\ &+ \sum_{m,n} b_{mn} J_m(\lambda_{mn} r)\sin(m\theta)\cos(\lambda_{mn} ct) \\ &+ \sum_n A_{on} J_0(\lambda_{on} r)\sin(\lambda_{on} ct) \\ &+ \sum_{m,n} A_{mn} J_m(\lambda_{mn} r)\cos(m\theta)\sin(\lambda_{mn} ct) \\ &+ \sum_{m,n} B_{mn} J_m(\lambda_{mn} r)\sin(m\theta)\sin(\lambda_{mn} ct), \end{aligned} \tag{12.12}$$

where J_m represents the m-th Bessel function of the first kind, λ_{mn} represents the n-th zero of J_m, In the subsequent we present the numerical solution of the Eq. (10.7 for the thermal wave with velocity $v = 5\ 10^{-3} c$, c = light velocity. Considering formula (10.9) for relaxation time we obtain

$$\tau = \frac{\hbar}{mv^2} = 160 \text{ as} \tag{12.13}$$

for $m = m_e$, $(= 0.51\,\mathrm{MeV})$ i.e. for electrons and for mean free path of the electrons in the 2D electron gas. From formula (10.14) we conclude that λ_{mfp} is of the order of the de Broglie'a wave length of the electron. It means that for 2D electron gas in quantum stadium the hyperbolic quantum thermal equation can be applied

REFERENCES

[1] M. Kozlowski, J. Marciak-Kozlowska, *Attoscience,* arXiv 0806.0165.
[2] J. Marciak-Kozlowska, M. Kozlowski, *From femtoto attoscience and beyond,* Nova Science Publishers, New York, USA, 2009.

Chapter 13

KLEIN-GORDON EQUATION WITH CASIMIR POTENTIAL

The contemporary nanoelectronic develops the NEMS and MEMS structures in which the distance between the parts is of the order of nanometers. As was shown in monograph: *Form quarks to bulk matter* the transport phenomena on the nanoscale depends on the second derivative in time. It is in contrast to the macroscale heat transport where the Fourier law (only the first derivative in time). The second derivative in time term describes the memory of the thermally excited medium. Considering the contemporary discussion of the role played by Casimir force in the NEMS and MEMS in this paper we describe the heat signaling in the simple nanostructure the parallel plates heated by attosecond laser pulses. It will be shown that temperature field between plates depends on the distance of the plates (in nanoscale). As the result the attosecond laser pulse can be used as the tool for the investigation of Casimir effect on the performance of the NEMS and MEMS.

REPULSIVE QUANTUM VACUUM FORCES

Vacuum energy is a consequence of the quantum nature of the electromagnetic field, which is composed of photons. A photon of frequency ω has energy $\hbar\omega$, where $\hbar$ is Planck constant. The quantum vacuum can be interpreted as the lowest energy state (or ground state) of the electromagnetic (EM) field that occurs when all charges and currents have been removed and the temperature has been reduced to absolute zero. In this state no ordinary photons are present. Nevertheless, because the electromagnetic field is a quantum system the energy of the ground state of the EM is not zero. Although the average value of the electric field $\langle E \rangle$ vanishes in ground state, the Root Mean Square of the field $\langle E^2 \rangle$ is not zero. Similarly the $\langle B^2 \rangle$ is not zero. Therefore the electromagnetic field energy $\langle E^2 \rangle + \langle B^2 \rangle$ is not equal zero. A detailed theoretical calculation tells that EM energy in each mode of oscillation with frequency ω is 0.5 $\hbar\omega$, which equals one half of amount energy that would be present if a single "real" photon of that mode were present. Adding up 0.5 $\hbar\omega$ for all possible modes of electromagnetic field gives a very large number for the vacuum energy E_0 in the quantum vacuum

$$E_0 = \sum_i \frac{1}{2}\hbar\omega_i. \tag{13.1}$$

The resulting vacuum energy E_0 is *infinity* unless a high frequency cut off is applied.

Inserting surfaces into the vacuum causes the modes of the EM to change. This change in the modes that are present occurs since the EM must meet the appropriate boundary conditions at each surface. Surface alter the modes of oscillation and therefore the surfaces alter the energy density corresponding to the lowest state of the EM field. In actual practice the change in E_0 is defined as follows

$$\Delta E_0 = E_0 - E_S \tag{13.2}$$

where E_0 is the energy in empty space and E_S is the energy in space with surfaces, i.e.

$$\Delta E_0 = \frac{1}{2}\sum_n^{\substack{\text{empty}\\\text{space}}}\hbar\omega_n - \frac{1}{2}\sum_i^{\substack{\text{surface}\\\text{present}}}\hbar\omega_i\,. \tag{13.3}$$

As an example let us consider a hollow conducting rectangular cavity with sides a_1, a_2, a_3. In that case for uncharged parallel plates with area A the attractive force between the plates is

$$F_{att} = -\frac{\pi^2\hbar c}{240d^4}A, \tag{13.4}$$

where d is the distance between plates. The force F_{att} is called the parallel plate Casimir force. Recent calculation show that for conductive rectangular cavities the vacuum forces on a given face can be repulsive (positive), attractive (negative) or zero depending on the ratio of the sides [1-2].

The first measurement of repulsive Casimir force was performed for the distance (separation) $d \sim 0.1$ μm the repulsive force is of the order of 0.5 μN – for cavity geometry. In March 2001, scientist at Lucent Technology used attractive parallel plate Casimir force to actuate a MEMS torsion device].

KLEIN – GORDON EQUATION WITH CASIMIR FORCE

Standard Klein – Gordon equation reads:

$$\frac{1}{c^2}\frac{\partial^2\Psi}{\partial t^2} - \frac{\partial^2\Psi}{\partial x^2} + \frac{m^2c^2}{\hbar^2}\Psi = 0. \tag{13.5}$$

In equation (5) Ψ is the relativistic wave function for particle with mass m, c is the light velocity and $\hbar$ is Planck constant. For massless particles $m = 0$ Eq. (13.5) is the Maxwell equation for photons. As was shown by Pauli and Weiskopf since relativistic quantum mechanical equation had to allow for creation and annihilation of particles, the Klein – Gordon describes spin – 0 bosons.

In monograph [2] the generalized Klein – Gordon thermal equation was developed

$$\frac{1}{v^2}\frac{\partial^2 T}{\partial t^2} - \nabla^2 T + \frac{m}{\hbar}\frac{\partial T}{\partial t} + \frac{2Vm}{\hbar^2} = 0. \tag{13.6}$$

In Eq. (13.6) T denotes temperature of the medium and v is the velocity of the temperature signal in the medium. When we extract the highly oscillating part of the temperature field,

$$T = e^{-\frac{t\omega}{2}} u(x,t), \tag{13.7}$$

where $\omega = \tau^{-1}$, and τ is the relaxation time, we obtain from Eq. (13.3) (1D case)

$$\frac{1}{v^2}\frac{\partial^2 u}{\partial t^2} - \frac{\partial^2 u}{\partial x^2} + qu(x,t) = 0, \tag{13.8}$$

where

$$q = \frac{2Vm}{\hbar^2} - \left(\frac{mv}{2\hbar}\right)^2. \tag{13.9}$$

When $q > 0$ equation (13.4) is of the form of the Klein – Gordon equation in the potential field $V(x, t)$. For $q < 0$ Eq. (13.8) is the modified Klein – Gordon equation. The discussion of the physical properties of the solution of equation (13.4) an be find in [1].

In the case of heat signaling in the medium excited by ultra-short laser pulses, $\Delta t < \tau$, the solution of Eq. (13.6) can be approximated as

$$T(x,t) \cong u(x,t) \qquad \text{for} \qquad \Delta t << \tau. \tag{13.10}$$

Considering the existence of the attosecond laser with Δt = 1 as = 10^{-18}s, Eq. (13.8) describes the heat signaling for thermal energy transport induced by ultra-short laser pulses. In the subsequent we will consider the heat transport when V is the Casimir potential. As was shown in monograph [2] the Casimir force, formula (13.4), can be repulsive sign (V) = +1 and attractive sign (V) = -1. For attractive Casimir force, $V < 0$, $q < 0$ (formula (13.5)) and equation (13.4) is the modified K – G equation. For repulsive Casimir force $V > 0$ and q can be positive or negative.. In subsequent we consider Cauchy initial condition:

$$u(x,0) = 0, \qquad \frac{\partial u(x,0)}{\partial t} = f(x)$$

The solution of Eq. (13.8) has the form

$$u(x,t) = \frac{1}{2v}\int_{x-vt}^{x+vt} f(x)J_0\sqrt{q\left(v^2t^2-(x-\zeta)^2\right)}d\zeta \qquad \text{for} \quad q>0 \tag{13.11}$$

and

$$u(x,t) = \frac{1}{2v}\int_{x-vt}^{x+vt} f(x)I_0\sqrt{-q\left(v^2t^2-(x-\zeta)^2\right)}d\zeta \qquad \text{for} \quad q>0 \tag{13.12}$$

REFERENCES

[1] M. Kozlowski, J. Marciak-Kozlowska, *Attoscience,* arXiv 0806.0165

[2] J. Marciak-Kozlowska, M. Kozlowski, *From femto-to attoscience and beyond,* Nova Science Publishers, New York, USA, 2009.

Chapter 14

RELATIVISTIC DESCRIPTION OF THE FIELD MATTER INTERACTION

When the electromagnetic field interacts with the medium the temperature of the medium depends on the velocity of the the medium/ observers. The relativistic formulation of thermodynamics was taken up already by Einstein himself and by several other physicist, notably Planck and von Laue. The principal result at this stage was the following

$$T = T^{O}\sqrt{1-\frac{\upsilon^2}{c^2}} \tag{14.1}$$

In this equation T is the absolute temperature in a system which is moving with a velocity υ with respect to the rest system (which is indicated by a superscript *0*)

In the paper by Ott [1] the traditional formulation (1) was questioned. Instead of the transformation (1) for the temperature T, Ott requires

$$T = \frac{T^{O}}{\sqrt{1-\frac{\upsilon^2}{c^2}}} \tag{14.2}$$

In this paper considering hyperbolic heat transport equation the temperature transformation equation will be obtained. It will be shown that the Ott formulation is valid Lorentz transformation for temperatures.

The model equation As was shown in monograph [2] the master equation for the heat transport induced by ultra-short laser pulses can be written as:

$$\vec{q} + \tau\frac{\partial \vec{q}}{\partial t} = -\kappa\frac{\partial T}{\partial x} \tag{14.3}$$

$$\frac{\partial \vec{q}}{\partial x} + c_V\frac{\partial T}{\partial t} = 0 \tag{14.4}$$

In Eq.(14.3) $\vec{q}$ is the heat current, T denotes temperature, τ is the relaxation time and κ is the heat conduction coefficient. In Eq.(14.4) c_V is the specific heat at constant volume. For quantum limit of the heat transport τ is equal

$$\tau = \frac{\hbar}{m\upsilon^2} \tag{14.5}$$

where m is the mass of heat carriers and $\upsilon = \alpha c$. The constant α_i is the coupling constant, $\alpha_1 = 1/137$ for electromagnetic interaction and $\alpha_2 = 0.15$ for strong interaction, c is the vacuum light speed.

From Eqs (14.1) and (14.2) hyperbolic diffusion equation for temperature can be derived

$$\tau\frac{\partial^2 T}{\partial t^2} + \frac{\partial T}{\partial t} = D\frac{\partial^2 T}{\partial x^2}, \tag{14.6}$$

where thermal diffusion coefficient, D,

$$D = \frac{\kappa}{c_V}.$$

The basic principle of the special relativity theory can be stated as:

All physics laws look the same in all inertial references frames. Equation (14.4) is the conservation of thermal energy. Let us consider two infinite rods K and K', where K' is moving with velocity v parallel to axis X. From the relativity principle we obtain

$$\begin{aligned} &\frac{\partial \vec{q}}{\partial x} + c_V\frac{\partial T}{\partial t} = 0 \text{ in frame } K \\ &\frac{\partial \vec{q}'}{\partial x'} + c_V\frac{\partial T'}{\partial t'} = 0 \text{ in frame } K' \end{aligned} \tag{14.7}$$

As can be easily shown the $\vec{q}$, $c_V T$ and $\vec{q}\,'$, $c_V T'$ are transformed according to Lorentz transformation:

$$\begin{aligned} q'_{x'} &= \gamma\left(q_x - \upsilon c_V T\right) \\ q'_{y'} &= q_y \\ q'_{z'} &= q_z \\ c_V T' &= \gamma\left[c_V T - \frac{\upsilon}{c^2}q_x\right] \end{aligned} \tag{14.8}$$

and

$$
\begin{aligned}
q_x &= \gamma(q'_{x'} + \upsilon c_V T') \\
q_y &= q'_{y'} \\
q_z &= q'_{z'} \\
c_V T &= \gamma\left[c_V T' + \frac{\upsilon}{c^2} q'_x\right]
\end{aligned}
\tag{14.9}
$$

The current q_x and temperature T form the four vectors.
The space time interval Δ,

$$
\Delta = q_x^2 - c^2 c_V T^2 = q'^2_{x'} - c^2 c_V T'^2 \tag{14.10}
$$

is invariant under the Lorentz transformations (14.8) and (14.9).

Let us consider the case where there is no heat current in rod K, i.e. when the rod has the temperature T = const and $\nabla T = 0$. In that case $q_x = 0$ and from formulae (14.8) we obtain

$$
\begin{aligned}
&q'_x = -\gamma\, \upsilon\, c_V T \\
&c_V T' = \gamma\, c_V T; \qquad T' = \gamma\, T \\
&\quad T' > T \quad \text{for} \quad \gamma > 1
\end{aligned}
\tag{14.11}
$$

Formula (14.11) is in agreement with Ott result. As the result we obtain that temperatures of the rods are different. And the moving rod observes the greater temperature $T' = \gamma T$.

From Eq.(14.10) we conclude that

$$
q'^2_{x'} = c^2 c_V \left(T'^2 - T^2\right) \tag{14.12}
$$

and $q^2_{x'} > 0, q_x' > 0$. It means that the heat current is directed parallel to the moving of the rod K in the reference frame K'.

REFERENCES

[1] H. Ott, *Z. Phys. 175* (1963) 70

[2] J. Marciak-Kozlowska, M. Kozlowski, *From femto-to attoscience and beyond,* Nova Science Publishers, New York, USA, 2009

Chapter 15

ON THE POSSIBLE FIELD GENERATED TACHYONS

The square of the neutrino mass was measured in tritium beta decay experiments by fitting the shape of the beta spectrum near endpoint. In many experiments it has been found to be negative. According to the results of paper [1]

$$m^2(\upsilon_e) = -2.5 \text{ eV}^2.$$

Based on special relativity superluminal particles, i.e. particles with $m^2 < 0$ and $\upsilon > c$ were proposed and discussed in monograph [2]. In this paper we investigate the possibility of the existence of the superluminal particles from the point of view the hyperbolic heat transport equation. It will be shown that hyperbolic heat transport equation is invariant under transformation $c^2 \rightarrow -c^2$, i.e. for transformation $c \rightarrow ic$. The new Lorentz transformation for $-c^2$ and formula for kinetic energy will be developed. It will be shown that for $\frac{E_k}{mc^2} > 1$ the speed of particles is decreasing for increased kinetic energy E_k.

The Equation In our monograph the hyperbolic Heaviside transport equation for attosecond laser pulses was obtained [2]

$$\frac{1}{\upsilon^2}\frac{\partial^2 T}{\partial t^2} + \frac{1}{D_T}\frac{\partial T}{\partial t} = \nabla^2 T. \tag{15.1}$$

In this equation T is the absolute temperature, υ denotes the speed of the thermal disturbance, and D_T is diffusion coefficient for thermal phenomena. Equation (1) describes the damped thermal wave propagation. Recently, the observation of the thermal wave in GaAs films exposed to ultra-short laser pulses was presented [2]. In the subsequent we will consider the mathematical structure of the Eq. (1) which can be interested for both the experimentalist as well as the theorist involved in ultrahigh ultra-short laser pulses. In the monograph [3] it was shown that speed υ, and diffusion coefficient D_T in Eq. (15.1) can be written as

$$\upsilon = \alpha c,$$

$$D = \frac{\hbar}{m}. \tag{15.2}$$

In formula (15.2) α is the electromagnetic fine structure constant, $\hbar$ is the Planck constant, m is the mass of the heat carrier, and c is light velocity.

Let us consider the transformation

$$c \to ic \tag{15.3}$$

for the Eq. (1). First of all we note that

$$\begin{aligned} v'^2 &\to \alpha^2 c^2 = v^2, \\ D'_T &= D_T. \end{aligned} \tag{15.4}$$

We conclude that the Eq. (15.1) is invariant under the transformation (15.3). One can say that Eq. (15.1) is valid for the universe for which $-c^2$ is the invariant constant.

THE LORENTZ TRANSFORMATION FOR $-c^2$ UNIVERSE

In our Universe the Lorentz transformation has the form

$$\begin{aligned} x' &= \frac{x - vt}{\sqrt{1 - \frac{v^2}{c^2}}}, \\ t' &= \frac{t - \frac{v}{c^2}x}{\sqrt{1 - \frac{v^2}{c^2}}}. \end{aligned} \tag{15.5}$$

With the transformation (15.3) we obtain from formula (15.5)

$$\begin{aligned} x' &= \frac{x - vt}{\sqrt{1 + \frac{v^2}{c^2}}}, \\ t' &= \frac{t + \frac{v}{c^2}x}{\sqrt{1 - \frac{v^2}{c^2}}}. \end{aligned} \tag{15.6}$$

For the space-time interval we obtain

$$x^2 + c^2 t^2 = x'^2 + c^2 t'^2$$

as in c^2 special relativity.

For $-c^2$ SR we have new formula for the velocities

$$v' = \frac{v - V}{1 + \frac{vV}{c^2}}. \tag{15.7}$$

For speed $v = ic$, from formula (15.7) we obtain $v' = ic$, i.e. object with speed ic has the same speed in all inertial reference frames. Now, we consider the formulae for total energy and momentum of the particle with mass m,

$$E = \frac{-mc^2}{\sqrt{1 + \frac{v^2}{c^2}}}, \qquad p = \frac{mv}{\sqrt{1 + \frac{v^2}{c^2}}}. \tag{15.8}$$

From formula (15.8) we obtain

$$\frac{E}{p} = -\frac{c^2}{v}. \tag{15.9}$$

For objects with speed $v = ic$ we obtain from formula (15.9)

$$\frac{E}{p} = ic. \tag{15.10}$$

Considering formula

$$E = \sqrt{-p^2 c^2 + m^2 c^4} \tag{15.11}$$

we obtain, for $m = 0$

$$E = ipc, \qquad \frac{E}{p} = ic. \tag{15.12}$$

We conclude that objects with masses $m = 0$, have speed $v = ic$ in universe. We calculate the kinetic energy, E_k

$$E_k = -mc^2(\gamma - 1), \qquad \gamma = \frac{1}{\sqrt{1 + \frac{v^2}{c^2}}}. \tag{15.13}$$

From formula (15.13) we deduce the ratio $\frac{v^2}{c^2}$.

$$\frac{v^2}{c^2} = \frac{1 - \left(1 - \frac{E_k}{mc^2}\right)^2}{\left(1 - \frac{E_k}{mc^2}\right)^2}. \tag{15.14}$$

It is quite interesting that in $-c^2$ universe $\frac{v^2}{c^2}$ is singular for $E_k = mc^2$, i.e. when kinetic energy of the objects equals its internal energy. For the c^2 universe the formula which describes $\frac{v^2}{c^2}$ reads

$$\frac{v^2}{c^2} = \frac{\left(1 + \frac{E_k}{mc^2}\right)^2 - 1}{\left(1 + \frac{E_k}{mc^2}\right)^2}. \tag{15.15}$$

In monograph [2] we show that the Eq. (15.1) describes the propagation of heatons, quanta of thermal field *T(x,t)*. For quantum heat transfer equation (15.1) we seek solution in the form

$$T(x,t) = e^{-\frac{t}{2\tau}} u(x,t). \tag{15.16}$$

After substitution of Eq. (15.16) into Eq. (15.1) one obtains

$$\frac{1}{v^2}\frac{\partial^2 u}{\partial t^2} - \frac{\partial^2 u}{\partial x^2} + qu(x,t) = 0, \tag{15.17}$$

where

$$q = -\left(\frac{mv}{2\hbar}\right)^2. \tag{15.18}$$

It is interesting to observe that if we introduce the imaginary mass $m^* = im$ then formula can be written as

$$q = -\left(\frac{m^* \upsilon}{2\hbar}\right)^2, \tag{15.19}$$

and Eq. (15.17) is the Klein-Gordon for *heatons* with imaginary mass $m^* = im$. According to the results of our monograph[3] we can call *heatons* with imaginary mass m^* the *tachyons,* i.e. particles with $\upsilon > c$.

On the experimental ground the existence of the *tachyons* is still an open question. However, the results of [1, 2] strongly suggest the existence of the particles with masses $m^* = im$, i.e. $(m^*)^2 = -m^2 < 0$.

REFERENCES

[1] T. Chang, *Nucl. Sci. Tech. 13,* (2002) 129.
[2] J. Marciak-Kozlowska, M. Kozlowski, *From femto-to attoscience and beyond,* Nova Science Publishers, New York, USA, 2009.

Chapter 16

On the Possible Void Decay

We must make some profound alterations to the theoretical idea of the vacuum. ...
Thus, with the new theory of electrodynamics we are rather forced to have an aether.
PAM Dirac, Nature, 1951, vol. 168, pp. 906-907

Recently the ultra-high energy lasers proposals are developed [1]. The SASE-FEL project for the first time enables the investigation of the "vacuum decay" processes and emission of ultra-relativistic fermions. In this paper we argue that the results of the SASEFEL future experiments opens the new field of the investigation of the structure of the spacetime.

The Model Description

Bell's theorem is rooted in two assumptions: the objective reality – the reality of the external world, independent of our observations; the second is locality, or no faster than light signaling. Aspect's experiment appear to indicate that one of these two has to go.

In this paper we are going back to relativity as it was before Einstein when people like Lorentz and Poincaré thought that there was an aether – a preferred frame of reference – but that our measuring instruments were distorted by motion in such a way that we could not detect motion through the aether. Now in that way you can imagine that there is a preferred frame of reference and in this preferred frame of reference particles do go faster than light. But then in other, our, frame of reference particles have the speed lower than the light speed.

In this paper we propose the following scenario. Behind the scene – our world of observation something is going which is not allowed to appear on the scenes.

To start with we observe that for electrons, at the vicinity of $E_k \sim mc^2$ the speed of electrons is changed abruptly. With $E_k \sim mc^2$, through Heisenberg inequality the characteristic time can be defined

$$\tau = \frac{\hbar}{mc^2}. \tag{16.1}$$

At that characteristic time the Newton theory (NT) and SR theory starts to give different description of the speeds, viz.,

$$v_{NT} = \sqrt{\frac{2E_k}{m_k}} \tag{16.2}$$

and

$$v_{SR} = c\sqrt{1 - \frac{1}{\left(\frac{E_k}{mc^2}+1\right)^2}}. \tag{16.3}$$

Let us introduce the acceleration a_m which describes the change of speeds in time τ

$$a_m = \frac{v_{NT} - v_{SR}}{\tau}. \tag{16.4}$$

and force F which opposes the acceleration of the particle with mass m

$$\begin{aligned} F = ma_m &= \frac{mc}{\tau}\left(\left(\frac{2E_k}{mc^2}\right)^{\frac{1}{2}} - \left(1 - \frac{1}{\left(\frac{E_k}{mc^2}+1\right)^2}\right)^{\frac{1}{2}}\right) \\ &= \frac{m^2c^3}{\hbar}\left(\left(\frac{2E_k}{mc^2}\right)^{\frac{1}{2}} - \left(1 - \frac{1}{\left(\frac{E_k}{mc^2}+1\right)^2}\right)^{\frac{1}{2}}\right). \end{aligned} \tag{16.5}$$

In formula (16.5) we introduce the field E_s

$$E_s = \frac{m^2c^3}{\hbar e} \tag{16.6}$$

and we obtain:

$$F = E_s e = \left(\frac{2E_k}{mc^2}\right)^{\frac{1}{2}} - \left(1 - \frac{1}{\left(\frac{E_k}{mc^2}+1\right)^2}\right)^{\frac{1}{2}}. \tag{16.7}$$

It is interesting to observe that the field E_s is the same as the Schwinger field strengths [2]. J. Schwinger demonstrated that in the background of a static electric field, the QED vacuum is broken and decayed with spontaneous emission of e^+e^- pairs.

In the following we define for electrons the energy L

$$L = eF_s r_e = \alpha \frac{(m_e c^2)^4}{e^6}. \tag{16.8}$$

In the formula (16.8) r_e is the classical electron radius

$$r_e = \frac{e^2}{m_e c^2} \tag{16.9}$$

and α is the fine structure constant. Having the energy L and volume r_e^3 where the energy is concentrated we define the bulk modulus for the medium which opposes the motion of electron

$$B = \frac{L}{r_e^3} = \alpha \frac{(m_e c^2)^4}{e^6} \tag{16.10}$$

and hypothetic sound velocity in the medium which oppose the acceleration of the electrons

$$\upsilon_{sound} = \left(\frac{B}{\rho}\right)^{\frac{1}{2}} = 10^{18} c, \tag{16.11}$$

where $\rho = \frac{m}{r_e^3}$.

REFERENCE

[1] TESLA, SASE-FEL *Technical Design Project.*

Chapter 17

RELATIVISTIC FIELD DESCRIPTION OF THE NANOTECHNOLOGY PHENOMENA

The nanotechnology industry has evolved at very high rate, particularly in the past ten years. transistor channel length have decreased from 2.0 um in 1980 to 0.5 um in 1992 to current (2014) systems with channel length ~ 50nm. Recently the International Technology Road Map for Semiconductors (ITRS) was postulated. The IRST is devoted to the study of the proposed of the global interconnects and problems with the time delays of the transistors and logical systems. One of the *remedium* for system time delays is the technology On –Chip Transmission Lines. The On-Chip transmission Technology allows for the electron transmission with near- of – light On- Chip electrical interconnection

This paragraph is addressed to the thermal transport in On-Chip –Transmission Lines. We develop the theoretical framework for heat transport in On-Chip- Transmission Line (OCTL) We formulate the transmission line heat transport equation and solved it for the Cauchy boundary conditions.

We will study the special relativity influence on the information transmission in OCTL. We show that the standard Fourier approximation leads to infinite speed of the data transmission. On the other hand the hyperbolic diffusion equation formulated in monograph[2] gives the finite speed of the transmission of the data.

THERMAL DIFFUSION IN OCTL

Dynamics of nonequilibrium electrons and phonons in metals, semiconductors have been the focus of much attention because of their fundamental interest in solid state physics and nanotechnology.

In metals, relaxation dynamics of optically excited nonequilibrium electrons has been extensively studied by pump – probe techniques using femtosecond lasers [1-2]

Recently [2] it was shown that the optically excited metals relax to equilibrium with two models: rapid electron relaxation and slow thermal relaxation through the creation of the optical phonons. The same processes will occur in OCTL

In this paper we develop the hyperbolic thermal diffusion equation with two models: electrons and phonons relaxation. These two modes are characterized by two relaxation times

τ_1 for electrons and τ_2 for phonons. This new equation is the generalization of our one mode hyperbolic equation, with only electrons degrees of freedom, $\tau 1$ [2]. The hyperbolic two mode equation is the analogous equation to Klein – Gordon equation and allows the heat propagation with finite speed.

As was shown in paper [1] for high frequency laser pulses the diffusion velocity exceeds that of light. This is not possible and merely demonstrates that Fourier equation is not really correct. Oliver Heaviside was well aware of this writing [3]:

All diffusion formulae (as in heat conduction) show instantaneous action to an infinite distances of a source, though only to an infinitesimal extent. To make the theory of heat diffusion be rational as well as practical some modification of the equation to remove the instantaneity, however little difference it may make quantatively, in general.

August 1876 saw the appearance in Philosophical Magazine] the paper which extended the mathematical understanding the diffusion (Fourier) equation.[3] O. Heaviside for the first time wrote the hyperbolic diffusion equation for the voltage *V(x, t).* Assuming a uniform resistance, capacitance and inductance per unit length, $k, c$and $s$respectively he arrived at:

$$\frac{\partial^2 V(x,t)}{\partial x^2} = sc\frac{\partial^2 V(x,t)}{\partial t^2} + kc\frac{\partial V(x,t)}{\partial t}. \tag{17.1}$$

The discussion of the broad sense of the Heaviside equation (17.1) can be find out, for example in our monograph [1], viz,

$$\tau^2 \frac{\partial^2 T}{\partial t^2} + \tau \frac{\partial T}{\partial t} + \frac{2V\tau}{\hbar} T = \tau \frac{\hbar}{m} \nabla^2 T. \tag{17.2}$$

Equation (17.2) is the heat transport equation for the transmission line. In Eq. (17.2) $T(\vec{r},t)$ denotes the temperature field, V is the external potential, m is the mass of heat carrier and τ is the relaxation time

$$\tau = \frac{\hbar}{mv^2}. \tag{17.3}$$

As can be seen from formulae (17.2) and (17.3) in hyperbolic diffusion equation the same relaxation time τ is assumed for both type of motion: wave and diffusion.

This can not be so obvious. For example let us consider the simpler harmonic oscillator equation:

$$m\frac{d^2 x}{dt^2} + kx + c\frac{dx}{dt} = 0. \tag{17.4}$$

Equation (4) can be written as

$$\tau^2 \frac{d^2x}{dt^2} + \tau \frac{dx}{dt} + x = 0, \tag{17.5}$$

where

$$\tau^2 = \frac{m}{k}, \qquad \tau = \frac{c}{k} \tag{17.6}$$

i.e.

$$c^2 = km. \tag{17.7}$$

As it was well known equation (17.4) with formula (17..6) describes only the weakly damped (periodic) motion of the harmonic oscillator (HO). It must be stressed that for HO exists also critically damped and overdamped modes which are not describes by the equation (17.5). The general master equation for HO must be written as

$$\tau_1^{\ 2} \frac{d^2x}{dt^2} + \tau_2 \frac{dx}{dt} + x = 0. \tag{17.8}$$

We argue that the general hyperbolic diffusion equation can be written as:

$$\tau_1^{\ 2} \frac{\partial^2 T}{\partial t^2} + \tau_2 \frac{\partial T}{\partial t} + \frac{2V\tau_2}{\hbar} T = \tau_2 \frac{\hbar}{m} \nabla^2 T \tag{17.9}$$

and $\tau_1 \neq \tau_2$.

Equation (17.9) describes the temperature field generated by ultra-short laser pulses. In Eq. (17.9) two modes: wave and diffusion are described by different relaxation times.

$$T(x,t) = e^{-\frac{t\tau_2}{2\tau_1^{\ 2}}} u(x,t). \tag{17.10}$$

After substitution Eq. (17.10) into Eq. (17.9) one obtains

$$\frac{1}{\upsilon^2} \frac{\partial^2 u}{\partial t^2} - \frac{\partial^2 u}{\partial x^2} + qu(x,t) = 0, \tag{17.11}$$

where

$$\upsilon^2 = \frac{\hbar \tau_2}{m\tau_1^{\ 2}}, \qquad q = \left(\frac{2Vm}{\hbar^2} - \frac{1}{4} \frac{m}{\hbar} \frac{\tau_2}{\tau_1^{\ 2}} \right). \tag{17.12}$$

Equation (17.11) is the thermal two-mode Klein – Gordon equation and is the generalization of Klein – Gordon one mode equation developed in our monograph.

For Cauchy initial condition

$$u(x,0) = f(x), \qquad u_t(x,0) = g(x) \tag{1713}$$

the solution of Eq. (17.11) has the form

$$\begin{aligned} u(x,t) = &\frac{f(x-\upsilon t)+f(x+\upsilon t)}{2} \\ &+\frac{1}{2\upsilon}\int_{x-\upsilon t}^{x+\upsilon t} g(\varsigma)I_0\left[\sqrt{-q\left(\upsilon^2t^2-(x-\varsigma)^2\right)}\right]d\varsigma \\ &+\frac{\upsilon\sqrt{-q}t}{2}\int_{x-\upsilon t}^{x+\upsilon t} f(\varsigma)\frac{I_1\left[\sqrt{-q\left(\upsilon^2t^2-(x-\varsigma)^2\right)}\right]}{\sqrt{\upsilon^2t^2-(x-\varsigma)^2}}d\varsigma \end{aligned} \tag{17.14}$$

for $q < 0$

and

$$\begin{aligned} u(x,t) = &\frac{f(x-\upsilon t)+f(x+\upsilon t)}{2} \\ &+\frac{1}{2\upsilon}\int_{x-\upsilon t}^{x+\upsilon t} g(\varsigma)J_0\left[\sqrt{q\left(\upsilon^2t^2-(x-\varsigma)^2\right)}\right]d\varsigma \\ &+\frac{\upsilon\sqrt{q}t}{2}\int_{x-\upsilon t}^{x+\upsilon t} f(\varsigma)\frac{J_1\left[\sqrt{q\left(\upsilon^2t^2-(x-\varsigma)^2\right)}\right]}{\sqrt{\upsilon^2t^2-(x-\varsigma)^2}}d\varsigma \end{aligned} \tag{17.15}$$

for $q > 0$.

FOURIER DIFFUSION EQUATION AND SPECIAL RELATIVITY THEORY

In monograph [1] the speed of the diffusion signals was calculated

$$\upsilon = \sqrt{2D\omega} \tag{17.16}$$

where

$$D = \frac{\hbar}{m} \tag{17.17}$$

and ω is the angular frequency of the laser pulses. Considering formula (17.16) and (17.17) one obtains

$$\upsilon = c\sqrt{2\frac{\hbar\omega}{mc^2}} \tag{17.18}$$

and $\upsilon \geq c$ for $\hbar\omega \geq mc^2$.

From formula (17.18) we conclude that for $\hbar\omega > mc^2$ the Fourier diffusion equation is in contradiction with special relativity theory and breaks the causality in transport phenomena.

Hyperbolic Diffusion and Special Relativity Theory

In monograph [1] the hyperbolic model of the heat transport phenomena was formulated. It was shown that the description of the ultrashort thermal energy transport needs the hyperbolic diffusion equation (one dimension transport)

$$\tau\frac{\partial^2 T}{\partial t^2} + \frac{\partial T}{\partial t} = D\frac{\partial^2 T}{\partial x^2}. \tag{17.19}$$

In the equation (17.19) $\tau = \dfrac{\hbar}{m\alpha^2 c^2}$ is the relaxation time, m = mass of the heat carrier, α is the coupling constant and c is the light speed in vacuum, $T(x,t)$ is the temperature field and $D = \hbar/m$.

In monograph [2] the speed of the thermal propagation υ was calculated

$$\upsilon = \frac{2\hbar}{m}\sqrt{-\frac{m}{2\hbar}\tau\omega^2 + \frac{m\omega}{2\hbar}(1+\tau^2\omega^2)^{1/2}}. \tag{17.20}$$

Considering that $\tau = \hbar/m\alpha^2 c^2$ formula (17.20) can be written as

$$\upsilon = \frac{2\hbar}{m}\sqrt{-\frac{m}{2\hbar}\frac{\hbar\omega^2}{mc^2\alpha^2} + \frac{m\omega}{2\hbar}(1+\frac{\hbar^2\omega^2}{m^2c^4\alpha^4})^{1/2}}. \tag{17.21}$$

For

$$\frac{\hbar\omega}{mc^2\alpha^2} < 1, \quad \frac{\hbar\omega}{mc^2} < 1 \tag{17.22}$$

one obtains from formula (17.21)

$$\upsilon = \sqrt{\frac{2\hbar}{m}\omega} . \tag{17.23}$$

Formally formula (17.23) is the same as formula (17.18) but considering inequality (17.21) we obtain

$$\upsilon = \sqrt{\frac{2\hbar\omega}{m}} = \sqrt{2}\alpha c < c \tag{17.24}$$

and causality is not broken.

For

$$\frac{\hbar\omega}{mc^2} > 1 ; \frac{\hbar\omega}{\alpha^2 mc^2} > 1 \tag{17.25}$$

we obtain from formula (17.21)

$$\upsilon = \alpha c, \upsilon < c. \tag{17.26}$$

We conclude that the hyperbolic diffusion equation (17.19) describes the thermal phenomena in accordance with special relativity theory and causality is not broken independently of laser beam energy.

When the amplitude of the *electromagnetic* beam approaches the critical electric field of quantum electrodynamics the vacuum becomes polarized and electron – positron pairs are created in vacuum On a distance equal to the Compton length, $\lambda\!\!\!^{-}_C = \hbar/m_e c$, the work of critical field on an electron is equal to the electron rest mass energy $m_e c^2$, i.e. $eE_{Sch}\lambda\!\!\!^{-}_C = m_e c^2$. The dimensionless parameter

$$\frac{E}{E_{Sch}} = \frac{e\hbar E}{m_e^2 c^3} \tag{17.27}$$

becomes equal to unity for electromagnetic wave intensity of the order of

$$I = \frac{c}{r_e \lambda\!\!\!^{-}_C{}^2} \frac{m_e c^2}{4\pi} \cong 4.7 \cdot 10^{29} \tfrac{W}{cm^2} , \tag{17.28}$$

where r_e is the classical electron radius. For such ultra high intensities the effects of nonlinear quantum electrodynamics plays a key role: laser beams excite virtual electron – positron pairs. As a result the vacuum acquires a finite electric and magnetic susceptibility which lead to the scattering of light by light. The cross section for the photon – photon interaction is given by:

$$\sigma_{\gamma\gamma\to\gamma\gamma} = \frac{973}{10125}\frac{\alpha^3}{\pi^2} r_e^2 \left(\frac{\hbar\omega}{m_e c^2}\right)^6, \tag{17.29}$$

for $\hbar\omega / m_e c^2 < 1$ and reaches its maximum, $\sigma_{\max} \approx 10^{-20} cm^2$ for $\hbar\omega \approx m_e c^2$ [2].

Considering formulae (18) and (19) we conclude that linear hyperbolic diffusion equation is valid only for the laser intensities $I \leq 10^{29}$ W/cm^2. for high intensities the nonlinear hyperbolic diffusion equation must be formulated and solved.

REFERENCES

[1] M. Kozlowski, J. Marciak-Kozlowska, *Attoscience,* arXiv 0806.0165.
[2] J. Marciak-Kozlowska, M. Kozlowski, *From femto-to attoscience and beyond,* Nova Science Publishers, New York, USA, 2009.
[3] O. Heaviside, *Electric Papers,* vol. 1, in Elibron Classics, New York, 2006.

Chapter 18

TIME DELAY IN FIELD INDUCED PHOTOEMISSION

At the beginning of the last century physics was revolutionized by the discovery of the photoelectric effect]. This was the birth of quantum mechanics. Even today, the excitation and photoemission of the electrons from atoms by light remains one of the most important phenomena of quantum physics. Until now, it was assumed that the electron is released by the atom without delay following of the photons. Recently, a team of physicists from the Laboratory far Attosecond Physics (LAP) of Max Planck Institute of Quantum Optics (MPQ) and Ludwig – Maximillians Universitat in Münich (LMU), lead by professor F. Krausz has ascertained that electrons found in different orbitals within the atoms of the noble – gas neon leave the atom only after a finite time. The physicists fired pulses of near - infrared laser light lasting less than four femtoseconds at the noble – gas atoms. The atoms were simultaneously hit by extreme ultraviolet pulses of less than 200 as (attosecond), liberating electrons from their atomic orbitals. The attosecond flashes ejected electrons either from the outer 2p – orbitals or from the inner 2s – orbitals of the atom. With the controlled field of the synchronized laser pulse serving as an "attosecond chronomatograph", the physicist then recorded when the excited electrons left the atom. The measurement shows that, despite their simultaneous excitation the electron left the atoms with a time offset of around 20 as.

In our monograph [1] we study the thermal processes generated by attosecond laser pulses. We show that the interaction of the laser pulses with atoms must be described by time delayed hyperbolic thermal equation. The time delayed hyperbolic thermal equations describe the thermal processes with memory. The term memory lasts for a few relaxation times.

It is interesting to observe that the relaxation time of atomic electromagnetic interaction is described by formula

$$\tau = \frac{\hbar}{m_e \alpha^2 c^2}$$

where m_e is the electron mass, α is the fine structure constant $\alpha = 1/137$. From formula for τ we calculate $\tau \approx 10$ as. This value can be compared to delay time $\Delta t \approx 20$ as was measured?

In this monograph [1] we 1 study the problem of the time evolution of the ultra-short thermal processes. The pillars of quantum theory: Schrödinger equation and Heisenberg uncertainty principle are the approximations to more fundamental theory, which is not formulated yet. It is therefore important to keep an open mind and to explore the various alternatives. The correct conclusion to draw is that quantum theory is merely a special case of a much wider physics – a physics in which non-local (superluminal) signaling is possible and in which the uncertainty principle can be violated.

In a letter to Pauli of 28 October 1926 Heisenberg states [2]:

I have for all that a hope in a later solution of more or less the following kind (but one should not say something like this aloud) that space and time are really only statistical concepts, such as, say temperature, pressure etc, in gas... I often try to get further in this direction, but until now it will not work.

In this paragraph we will discuss the question of the local and nonlocal quantum thermal processes. To that aim we will consider the structure of the evolution equation for the temperature field $T(x,t)$ in 1D case.

Let us consider the hyperbolic quantum equation for heat transfer [1]:

$$\frac{1}{\upsilon^2}\frac{\partial^2 T(x,t)}{\partial t^2}+\frac{m}{\hbar}\frac{\partial T}{\partial t}+\frac{2Vm}{\hbar^2}T=\frac{\partial^2 T(x,t)}{\partial x^2}. \tag{18.1}$$

In formula (18.1) $T(x,t)$ is the temperature field, υ is the thermal signal velocity, m – mass of the heat carrier, V is the potential energy in which heat energy is transported.

In the following we will look for the short time, t, history of temperature propagation:

$$T(x,t)=e^{-\frac{t}{2\tau}}u(x,t), \tag{18.2}$$

where τ is the relaxation time for thermal processes.

Substituting formula (18.2) to the Eq.(18.1) we obtain the Klein – Gordon equation for $u(x,t)$:

$$\frac{1}{\upsilon^2}\frac{\partial^2 u(x,t)}{\partial t^2}-\frac{\partial^2 u(x,t)}{\partial x^2}+qu(x,t)=0. \tag{18.3}$$

where

$$q=\frac{2Vm}{\hbar^2}-\left(\frac{m\upsilon}{2\hbar}\right)^2 \tag{18.4}$$

Considering that [1]

$$\tau=\frac{\hbar}{m\upsilon^2}. \tag{18.5}$$

formula (2.4) can be written as

$$q = \frac{2Vm}{\hbar^2} - \left(\frac{1}{2\tau\upsilon}\right)^2. \tag{18.6}$$

From formula (18..6) we conclude that solution of the equation (2.1) for ultrashort time processes has three branches:

$$q > 0 \qquad V\tau > \hbar \tag{18.7}$$

$$q < 0 \qquad V\tau < \hbar \tag{18..8}$$

$$q = 0 \qquad V\tau = \hbar. \tag{18.9}$$

The first branch (18..7) is the quantum branch – Heisenberg uncertainty is fulfilled. The second branch (18.8) – Heisenberg uncertainty is not fulfilled and border branch.

As can be easily seen from formula (18.1) hyperbolic thermal equation is local quantum equation with $\upsilon=\alpha c$ finite. The parabolic thermal equation

$$\frac{m}{\hbar}\frac{\partial T}{\partial t} + \frac{2Vm}{\hbar^2}T = \frac{\partial^2 T(x,t)}{\partial x^2}. \tag{18..10}$$

is the nonlocal quantum diffusion equation for it can be obtained from equation (18..1) with $\upsilon \to \infty$, i.e. $c \to \infty$.

For quantum thermal processes standard equation is the Schrödinger equation which is the parabolic equation, i.e. is the non-local equation $c \to \infty$ [1].

Local and Nonlocal Description of the Attosecond Photoemission

The Compton formula for the photoemission is based on the Schrödinger equation for which $c \to \infty$ and relaxation time τ, formula (18.5) is going to zero. It means that in Compton formula the interaction time ≈ relaxation time is equal zero and photon – electron scattering is the instantenous phenomenon.

Let us look for the solution of the hyperbolic equation (18.11) for the Cauchy initial condition:

$$T(x,0) = 0, \qquad T(0,t) = f(t). \tag{18.11}$$

For initial condition (18.11) the solution of (18.1) has the form [1]:

$$T(x,t) = \left\{ f\left(t-\frac{x}{\upsilon}\right)e^{-\frac{\rho x}{\upsilon}} + \frac{\rho x}{\upsilon}\int_{\frac{x}{\upsilon}}^{t} f(t-y)e^{-y\rho}\frac{I_1\left[\rho\left(y^2-\frac{x^2}{\upsilon^2}\right)\right]}{\left(y^2-\frac{x^2}{\upsilon^2}\right)}dy \right\} H\left(t-\frac{x}{\upsilon}\right) \tag{18.12}$$

In formula (3.2) $\rho = \frac{1}{2\tau}$, $\tau = \frac{\hbar}{m\upsilon^2}$ and $H\left(t-\frac{x}{\upsilon}\right)$ is the unistep function

$$H\left(t-\frac{x}{\upsilon}\right) = 1 \quad \text{for} \quad t > \frac{x}{\upsilon}$$
$$H\left(t-\frac{x}{\upsilon}\right) = 0 \quad \text{for} \quad t = \frac{x}{\upsilon} \tag{18.13}$$

From formulae (18.12) and (18.13) we conclude that in the interaction of the laser pulse with atoms we have the delay described by the Heaviside step function, which operates only if we have the local interaction, i.e. υ is finite. For nonlocal interaction as in the case parabolic heat transport equation and Schrödinger equation $H\left(t-\frac{x}{\upsilon}\right) = H(t)$ in all heating region. The seminal experimental results of the LAP – the time delay of the photoemission opens the new field of experimental as well as theoretical investigations – the attosecond dynamics or for short attodynamics. The dynamics at the very short time periods is quite different from the description offered by standard quantum theory. In attosecond time window the memory of system: atom and incoming photon can not be overlooked.

REFERENCES

[1] M. Kozłowski, J. Marciak-Kozłowska, *Thermal processes using attosecond laser pulses*, Springer, 2006.

[2] W. Pauli, *Wisserschfliher Briefwechsel mit Bohr,* Einstein, Heisenberg u.a. Teil I, 1919 – 1929 ed. A. Herman et al. Springer 1979.

Part II. Strong Field

Chapter 19

NUCLEAR COLLECTIVE PROCESSES

With attosecond electromagnetic beams(1 as = 10^{-18} s) physicist and engineers are now close to controlling the motion of electrons on a timescale that is substantially shorter than the oscillation period of visible light. It is also possible with attosecond laser pulse to rip an electron wave-packet from the core of an atom and set it free with similar temporal precision.

Recently a laser configuration in which attosecond electron wave packets are ionized and accelerated to multi-MeV energies, was proposed [1]. This technique opens an avenue towards imaging attosecond dynamics of nuclear processes.

High laser intensity atomic and molecular physics is dominated by the recollision between and ionized electron and its parent ion. The electron is ionized near the peak of laser field, accelerated away from the ion and driven back to its parent ion once the field direction reverses. Recollision leads to nonsequential double ionization, high harmonic generation, and attosecond extreme ultraviolet and electron pulses [1].

In contrast the extension of laser induced recollision physics to relativistic energies is a long standing issue. The solution to this problem was achieved in paper [1]. The authors of the paper [1] show that the Lorentz force (which prevents recollision for relativistic electrons) is eliminated for two counterpropagating, equally handed, circularly polarized beams through the whole focal volume as long as the laser pulses are sufficiently long. In this configuration, the recollision energy is limited only by the maximum achievable laser intensity (currently ~ 10^{23} W/cm^3).

The method presented in paper [1] will alter the physics anywhere relativistic laser fields interact with electrons. For example the attosecond laser pulses can resolve nuclear dynamics. In nuclear physics it opens novel possibilities in the study of decay and damping of nuclear dynamical processes such as giant resonances. It reveals information on fundamental nuclear properties, such as dissipation and viscosity in nuclei.

Relativistic hyperbolic heat transport equation for nuclear processes

In paper [3] relativistic hyperbolic transport equation (RHT) was formulated:

$$\frac{1}{\upsilon^2}\frac{\partial^2 T}{\partial t^2}+\frac{m_0\gamma}{\hbar}\frac{\partial T}{\partial t}=\nabla^2 T. \tag{19.1}$$

In equation (19.1) υ is the velocity of heat waves, m_0 is the mass of heat carrier (nucleon)and γ – the Lorentz factor $\gamma = \left(1 - \frac{\upsilon^2}{c^2}\right)^{-1/2}$. As was shown in paper [1] the heat energy (*heaton temperature*) T_h can be defined as follows:

$$T_h = m_0 \gamma \upsilon^2 . \tag{19.2}$$

Considering that υ, thermal wave velocity equals [1]

$$\upsilon = \alpha c , \tag{19.3}$$

where α is the coupling constant for the interactions which generate the *thermal wave* ($\alpha = 0.15$ for strong force) *heaton temperature* equals

$$T_h = \frac{m_0 \alpha^2 c^2}{\sqrt{1 - \alpha^2}} . \tag{19.4}$$

From formula (19.4) one concludes that *heaton temperature* is the linear function of the mass m_0 of the heat carrier. It is quite interesting to observe that the proportionality of T_h and heat carrier mass m_0 was the first time observed in ultrahigh energy heavy ion reactions measured at CERN -] it was shown that temperature of pions, kaons and protons produced in Pb+Pb, S+S reactions are proportional to the mass of particles. Recently at Rutherford Appleton Laboratory (RAL) the VULCAN laser was used to produce the elementary particles: electrons and pions

In the present paper the forced relativistic heat transport equation will be studied and solved. In monograph [1] the damped thermal wave equation was developed:

$$\frac{1}{\upsilon^2}\frac{\partial^2 T}{\partial t^2} + \frac{m}{\hbar}\frac{\partial T}{\partial t} + \frac{2Vm}{\hbar^2} T - \nabla^2 T = 0 . \tag{19.5}$$

The relativistic generalization of equation (19.5) is quite obvious:

$$\frac{1}{\upsilon^2}\frac{\partial^2 T}{\partial t^2} + \frac{m_0 \gamma}{\hbar}\frac{\partial T}{\partial t} + \frac{2Vm_0 \gamma}{\hbar^2} T - \nabla^2 T = 0 . \tag{19.6}$$

It is worthwhile to note that in order to obtain nonrelativistic equation we put $\gamma = 1$.

The motion of charged nucleons in the nucleus is equivalent to the flow of an electric current in a loop of wire. With attosecond laser pulses we will be able to influence the current in the nucleon "wire". This opens quite new perspective for the attosecond nuclear physics.

The new equation (19.6) is the natural candidate for the master equation which can be used to the description of heat transport in nuclear matter.

When external force is present $F(x,t)$ the forced damped heat transport is obtained instead of equation (19.6) (in one dimensional case):

$$\frac{1}{\upsilon^2}\frac{\partial^2 T}{\partial t^2}+\frac{m_0\gamma}{\hbar}\frac{\partial T}{\partial t}+\frac{2Vm_0\gamma}{\hbar^2}T-\frac{\partial^2 T}{\partial x^2}=F(x,t). \tag{19.7}$$

The hyperbolic relativistic quantum heat transport equation (RQHT), formula (19.7), describes the forced motion of heat carriers which undergo the scatterings $\left(\frac{m_0\gamma}{\hbar}\frac{\partial T}{\partial t}\text{ term}\right)$ and are influenced by potential $\left(\frac{2Vm_0\gamma}{\hbar^2}T\text{ term}\right)$.

The solution of equation can be written as

$$T(x,t)=e^{-\frac{t}{2\tau}}u(x,t), \tag{19.8}$$

where $\tau=\frac{\hbar}{\left(m\upsilon^2\right)}$ is the relaxation time. After substituting formula (19.8) to the equation (19.7) we obtain new equation

$$\frac{1}{\upsilon^2}\frac{\partial^2 u}{\partial t^2}-\frac{\partial^2 u}{\partial x^2}+qu(x,t)=e^{\frac{t}{2\tau}}F(x,t), \tag{19.9}$$

and

$$q=\frac{2Vm}{\hbar^2}-\left(\frac{m\upsilon}{2\hbar}\right)^2, \tag{19.10}$$

$$m=m_0\gamma.$$

Equation (19.9) can be written as:

$$\frac{\partial^2 u}{\partial t^2}-\upsilon^2\frac{\partial^2 u}{\partial x^2}+q\upsilon^2 u(x,t)=G(x,t), \tag{19.11}$$

where

$$G(x,t)=\upsilon^2 e^{\frac{t}{2\tau}}F(x,t).$$

When $q > 0$ equation (12.11) is the forced Klein-Gordon (K-G) equation. The solution of the forced Klein-Gordon equation for the initial conditions:

$$u(x,0) = f(x), \quad u_t(x,0) = g(z) \tag{19.12}$$

has the form [1]:

$$\begin{aligned} u(x,t) &= \frac{f(x-vt)+f(x+vt)}{2} \\ &+ \frac{1}{2v}\int_{x-vt}^{x+vt} g(\zeta)J_0\left[q\sqrt{v^2t^2-(x-\zeta)^2}\right]d\zeta \\ &- \frac{\sqrt{q}vt}{2}\int_{x-vt}^{x+vt} f(\zeta)\frac{J_1\left[q\sqrt{v^2t^2-(x-\zeta)^2}\right]}{\sqrt{v^2t^2-(x-\zeta)^2}}d\zeta \\ &+ \frac{1}{2v}\int_0^{t'}\int_{x-v(t-t')}^{x+v(t-t')} G(\zeta,t')J_0\left[q\sqrt{v^2(t-t')^2-(x-\zeta)^2}\right]d\zeta dt'. \end{aligned} \tag{19.13}$$

When $q < 0$ equation (12.11) is the forced modified Heaviside (telegraph) equation with the solution

$$\begin{aligned} u(x,t) &= \frac{f(x-vt)+f(x+vt)}{2} \\ &+ \frac{1}{2v}\int_{x-vt}^{x+vt} g(\zeta)J_0\left[-q\sqrt{v^2t^2-(x-\zeta)^2}\right]d\zeta \\ &- \frac{\sqrt{-q}vt}{2}\int_{x-vt}^{x+vt} f(\zeta)\frac{J_1\left[-q\sqrt{v^2t^2-(x-\zeta)^2}\right]}{\sqrt{v^2t^2-(x-\zeta)^2}}d\zeta \\ &+ \frac{1}{2v}\int_0^{t'}\int_{x-v(t-t')}^{x+v(t-t')} G(\zeta,t')J_0\left[-q\sqrt{v^2(t-t')^2-(x-\zeta)^2}\right]d\zeta dt'. \end{aligned} \tag{19.14}$$

When $q = 0$ equation (19.9)) is the forced thermal equation

$$\frac{\partial^2 u}{\partial t^2} - v^2\frac{\partial^2 u}{\partial x^2} = G(x,t). \tag{19.15}$$

On the other hand one can say that equation ((19.15) is the distortionless hyperbolic equation. The condition $q = 0$ can be rewrite as:

$$V\tau = \frac{\hbar}{8}. \tag{19.16}$$

The equation (19.16) is the analogous to the Heisenberg uncertainty relations

Subsequently we consider the initial and boundary value problem for the inhomogenous thermal wave equation in semi-infinite interval [1]: that is

$$\frac{\partial^2 u}{\partial t^2} - \upsilon^2 \frac{\partial^2 u}{\partial x^2} = G(x,t), \quad 0 < x < \infty, \quad t > 0, \tag{19.17}$$

with initial condition:

$$u(x,0) = f(x), \quad \frac{\partial u(x,0)}{\partial t} = g(x), \quad 0 < x < \infty,$$

and boundary condition

$$au(0,t) - b\frac{\partial u(0,t)}{\partial x} = B(t), \quad t > 0, \tag{19.18}$$

where $a \geq 0, b \geq 0, a + b > 0$ (with a and b both equal to constants) and F, f, g and B are given functions. The solution of equation (19.17, 19.18) is of the form [1]

$$\begin{aligned} u(x,t) &= \frac{1}{2}\left[f(x-\upsilon t) + f(x+\upsilon t)\right] \\ &+ \frac{1}{2\upsilon}\int_{x-\upsilon t}^{x+\upsilon t} g(s)ds \\ &+ \frac{1}{2\upsilon}\int_0^t \int_{x-\upsilon(t-t')}^{x+\upsilon(t-t')} F(s,t')dsdt' \end{aligned} \tag{19.19}$$

In the special case where $f = g = F = 0$ we obtain the following solution of the initial and boundary value problem (19.18)

$$u(x,t) = \begin{cases} 0, & x > \upsilon t, \\ \dfrac{\upsilon}{b}\displaystyle\int_0^{t-\frac{x}{\upsilon}} \exp\left[\frac{\upsilon a}{b}\left(y - t + \frac{x}{\upsilon}\right)\right] B(t)dt, & 0 < x < \upsilon t, \end{cases} \tag{19.20}$$

if $b \neq 0$. If $a = 0$ and $b = 1$, we have:

$$u(x,t) = \begin{cases} 0, & x > \upsilon t, \\ \upsilon \int_0^{t-\frac{x}{\upsilon}} B(y)dy, & 0 < x < \upsilon t. \end{cases}$$

It can be concluded that the boundary condition (19.18) gives rise to a wave of the form $K\left(t - \frac{x}{\upsilon}\right)$ that travels to the right with speed υ. For this reason the foregoing problem is often referred to as a *signaling problem* for the thermal waves.

As was shown in paper [1] the master equation for collective (thermal wave) motion of nucleons has the form

$$\frac{\partial^2 u}{\partial t^2} - \upsilon^2 \frac{\partial^2 u}{\partial x^2} + q\upsilon^2 u(x,t) = G(x,t). \tag{19.21}$$

When we are looking for undistorted motion, $q \to 0$ i.e.:

$$\frac{2Vm}{\hbar^2} - \left(\frac{m\upsilon}{2\hbar}\right)^2 \to 0.$$

For $q = 0$ one obtain

$$V\tau = \frac{\hbar}{8}. \tag{19.22}$$

For nuclear collective motion $\tau = \frac{\hbar}{m\upsilon^2}$, m is the nucleon mass and $\upsilon = \alpha_s c$, where $\alpha_s = 0.15$ is the strong coupling constant.

With the $m = 981\frac{\text{MeV}}{c^2}$ one obtains from formula (19.23)

$$V = \frac{\hbar}{8\tau} \approx 30\text{MeV}. \tag{19.23}$$

As it is well known the 30 MeV energy range is the location of giant dipole resonances in nuclear matter [1]. As was shown in paper [1] for attosecond laser induced recollision of electrons with ions the energy of the electrons is of the order of 30 MeV for $Z \approx 20$ (where Z is the atomic number of the ion). It means that with attosecond laser pulses the electronuclear giant resonances can be excited and investigated. structure constant for electromagnetic interaction, strong interaction and strong quark-quark interaction respectively, τ_i is the relaxation time for scattering process.

When the QHT is applied to the study of the thermal excitation of the matter, the quanta of thermal energy, the heaton can be defined with energies E_h^e = 9 eV, E_h^H = 7 MeV and E_h^q = 139 MeV for atomic, nucleon and quark level respectively.

REFERENCE

[1] J. Marciak-Kozlowska, M. Kozlowski arXiv 0403073.

Chapter 20

DOES FREE QUARKS EXIST?

One of the best models in mathematical physics is Fourier's model for the heat conduction in matter. Despite the excellent agreement obtained between theory and experiment, the Fourier model contains several inconsistent implications. The most important is that the model implies an infinite speed of propagation for heat. Cattaneo was the first to propose a remedy. He formulated new hyperbolic heat diffusion equation for propagation of the heat waves with finite velocity.

There is an impressive amount of literature on hyperbolic heat trans-port in matter paper [1] we developed the new hyperbolic heat transport equation which generalizes the Fourier heat transport equation for the rapid thermal processes. The hyperbolic heat conduction equation (HHC) for the fermionic system can be written in the form:

$$\frac{1}{\left(\frac{1}{3}\upsilon_F^2\right)}\frac{\partial^2 T}{\partial t^2}+\frac{1}{\tau\left(\frac{1}{3}\upsilon_F^2\right)}\frac{\partial T}{\partial t}=\nabla^2 T\,, \tag{20.1}$$

where T denotes the temperature, τ – the relaxation time for the thermal disturbance of the fermionic system and υ_F is the Fermi velocity.

In the subsequent we develop the new formulation of the HHC considering the details of the two fermionic systems: electron gas in metals and nucleon gas.

For the electron gas in metals the Fermi energy has the form:

$$E_F^e=(3\pi)^2\frac{n^{2/3}\hbar^2}{2m_e}, \tag{20.2}$$

where n – density and m_e – electron mass. Considering that

$$n^{-1/3}\sim a_B\sim\frac{\hbar^2}{me^2}, \tag{20.3}$$

and a_B – Bohr radius, one obtains

$$E_F^e \sim \frac{n^{2/3}\hbar^2}{2m_e} \sim \frac{\hbar^2}{ma^2} \sim \alpha^2 m_e c^2, \tag{20.4}$$

where c – light velocity and α = 1/137 is the fine structure constant. For the Fermi momentum, p_F we have

$$p_F^e \sim \frac{\hbar}{a_B} \sim \alpha m_e c, \tag{20.5}$$

and for Fermi velocity, υ_F

$$\upsilon_F^e \sim \frac{p_F}{m_e} \sim \alpha c.$$

Considering formula (20.5) equation (20.4) can be written as

$$\frac{1}{c^2}\frac{\partial^2 T}{\partial t^2} + \frac{1}{c^2 \tau}\frac{\partial T}{\partial t} = \frac{\alpha^2}{3}\nabla^2 T. \tag{20.6}$$

As it is seen from (20.4) the HHC equation is the relativistic equation as it takes into account the finite velocity of light. In order to derive the Fourier law from equation (8) we are forced to break the special theory of relativity and put in equation (20.6) $c \to \infty,\ \tau \to 0$. In addition it was demonstrated from HHC in a natural way, that in electron gas the heat propagation velocity $\upsilon_h \sim \upsilon_F$ in the accordance with the results of the pump probe experiments.

For the nucleon gas, Fermi energy is equal:

$$E_F^N = \frac{(9\pi)^{2/3}\hbar^2}{8mr_0^2}, \tag{20.7}$$

where m – nucleon mass and r_0, which describes the range of strong interaction is equal:

$$r_0 = \frac{\hbar}{m_\pi c}, \tag{20.8}$$

m_π – is the pion mass. Considering formula (20.8) one obtains for the nucleon Fermi energy

$$E_F^N \sim \left(\frac{m_\pi}{m}\right)^2 mc^2. \tag{20.9}$$

In the analogy to the equation (20.7) formula (20.9) can be written as follows

$$E_F^N \sim \alpha_s^2 mc^2, \tag{20.10}$$

where $\alpha_s = 0.15$ is fine structure constant for strong interactions. Analogously we obtain for the nucleon Fermi momentum

$$p_F^e \sim \frac{\hbar}{r_0} \sim \alpha_s mc \tag{20.11}$$

and for nucleon Fermi velocity

$$\upsilon_F^N \sim \frac{pF}{m} \sim \alpha_s c, \tag{20.12}$$

and HHC for nucleon gas can be written as follows:

$$\frac{1}{c^2}\frac{\partial^2 T}{\partial t^2} + \frac{1}{c^2 \tau}\frac{\partial T}{\partial t} = \frac{\alpha_s^2}{3}\nabla^2 T. \tag{20.13}$$

In the following the procedure for the discretization of temperature $T\,(r,\,t)$ in hot fermion gas will be developed. First of all we introduce the reduced de Broglie wavelength

$$\begin{aligned} \lambda_B^e &= \frac{\hbar}{m_e \upsilon_h^e}, \quad \upsilon_h^e = \frac{1}{\sqrt{3}}\alpha c, \\ \lambda_B^N &= \frac{\hbar}{m \upsilon_h^N}, \quad \upsilon_h^N = \frac{1}{\sqrt{3}}\alpha_s c, \end{aligned} \tag{20.14}$$

and mean free path, λ^e and λ^N

$$\lambda^e = \upsilon_h^e \tau^e, \quad \lambda^N = \upsilon_h^N \tau^N. \tag{20.15}$$

Considering formulas (20.14) and (20.15) we obtain HHC for electron and nucleon gases:

$$\frac{\lambda_B^e}{\upsilon_h^e}\frac{\partial^2 T}{\partial t^2} + \frac{\lambda_B^e}{\lambda^e}\frac{\partial T}{\partial t} = \frac{\hbar}{m_e}\nabla^2 T^e, \tag{20.16}$$

$$\frac{\lambda_B^N}{\upsilon_h^N}\frac{\partial^2 T}{\partial t^2} + \frac{\lambda_B^N}{\lambda^N}\frac{\partial T}{\partial t} = \frac{\hbar}{m}\nabla^2 T^N. \tag{20.17}$$

Equations (20.16, 20.17) are the hyperbolic partial differential equations which are the master equations for heat propagation in Fermi electron and nucleon gases. In the following we will study the quantum limit of heat transport in the fermionic systems. We define the quantum heat transport limit as follows:

$$\lambda^e = \lambdabar_B^e, \quad \lambda^N = \lambdabar_B^N. \tag{20.18}$$

In that case equations (20.16, 20.17) have the form:

$$\tau^e \frac{\partial^2 T^e}{\partial t^2} + \frac{\partial T^e}{\partial t} = \frac{\hbar}{m_e}\nabla^2 T^e, \tag{20.19}$$

$$\tau^N \frac{\partial^2 T^N}{\partial t^2} + \frac{\partial T^N}{\partial t} = \frac{\hbar}{m}\nabla^2 T^N, \tag{20.20}$$

where

$$\tau^e = \frac{\hbar}{m_e\left(\upsilon_h^e\right)^2}, \quad \tau^N = \frac{\hbar}{m\left(\upsilon_h^N\right)^2}. \tag{20.21}$$

Equations (20.19) and (20.20) define the master equation for quantum heat transport (QHT). Having the relaxation times τ^e and τ^N, one can define the "pulsations" ω_h^e and ω_h^N

$$\omega_h^e = (\tau^e)^{-1}, \quad \omega_h^N = (\tau^N)^{-1}, \tag{20.22}$$

or

$$\omega_h^e = \frac{m_e\left(\upsilon_h^e\right)^2}{\hbar}, \quad \omega_h^N = \frac{m\left(\upsilon_h^N\right)^2}{\hbar},$$

i.e.,

$$\begin{aligned} \omega_h^e\hbar &= m_e\left(\upsilon_h^e\right)^2 = \frac{m_e\alpha^2}{3}c^2, \\ \omega_h^N\hbar &= m\left(\upsilon_h^N\right)^2 = \frac{m\alpha_s^2}{3}c^2. \end{aligned} \tag{20.23}$$

The formulas (20.23) define the Planck-Einstein relation for heat quanta E_h^e and E_h^N

$$E_h^e = \omega_h^e \hbar = m_e \left(\upsilon_h^e\right)^2,$$
$$E_h^N = \omega_h^N \hbar = m_N \left(\upsilon_h^N\right)^2. \qquad (20.24)$$

The heat quantum with energy $E_h = \hbar\omega$ can be named the *heaton*, in complete analogy to the *phonon*, *magnon*, *roton*, etc. For $\tau^e, \tau^N \to 0$, Eqs. (20.72) and (20.76) are the Fourier equations with quantum diffusion coefficients D^e and D^N

$$\frac{\partial T^e}{\partial t} = D^e \nabla^2 T^e, \; D^e = \frac{\hbar}{m_e}, \qquad (20.25)$$

$$\frac{\partial T^N}{\partial t} = D^N \nabla^2 T^N, \; D^N = \frac{\hbar}{m}. \qquad (20.26)$$

The quantum diffusion coefficients D^e, D^N for the first time were introduced by E. Nelson

For finite τ^e and τ^N, for $\Delta t < \tau^e$, $\Delta t < \tau^N$, Eqs. (20.19) and (20.20) can be written as follows

$$\frac{1}{(\upsilon_h^e)^2}\frac{\partial^2 T^e}{\partial t^2} = \nabla^2 T^e, \qquad (20.27)$$

$$\frac{1}{(\upsilon_h^N)^2}\frac{\partial^2 T^N}{\partial t^2} = \nabla^2 T^N. \qquad (20.28)$$

Equations (20.27, 20.28) are the wave equations for quantum heat transport (QHT). For $\Delta t > \tau$, one obtains the Fourier equations

The possible interpretation of the heaton energies First of all we consider the electron and nucleon gases. For electron gas we obtain from formula (17), (26) for $m_e = 0.51$ MeV/c^2, $\upsilon_h = \alpha c / \sqrt{3}$

$$E_h^e = 9\,\text{eV}, \qquad (20.29)$$

which is of the order of the Rydberg energy. For nucleon gases one obtains ($m = 938$ MeV/c^2, $\alpha_s = 0.15$) from formulae (20.22)

$$E_h^N \sim 7\,\text{MeV} \qquad (20.30)$$

i.e. average binding energy of the nucleon in the nucleus ("boiling" temperature for the nucleus) When the ordinary matter (on the atomic level) or nuclear matter (on the nucleus level) is excited with short temperature pulses ($\Delta t \sim \tau$) the response of the matter is discrete. The matter absorbs the thermal energy in the form of the quanta E_h^e or E_h^N.

It is quite natural to pursue the study of the thermal excitation to the subnucleon level i.e. quark matter. In the following we generalize the QHT equation (8) for quark gas in the form:

$$\frac{1}{c^2}\frac{\partial^2 T^q}{\partial t^2}+\frac{1}{c^2\tau}\frac{\partial T^q}{\partial t}=\frac{\left(\alpha_s^q\right)^2}{3}\nabla^2 T^q \tag{20.31}$$

with α_s^q – the fine structure constant for strong quark-quark interaction and υ_s^q - thermal velocity:

$$\upsilon_h^q=\frac{1}{\sqrt{3}}\alpha_s^q c. \tag{20.32}$$

Analogously as for electron and nucleon gases we obtain for quark heaton

$$E_h^q=\frac{m}{3}\left(\alpha_s^q\right)^2 c^2, \tag{20.33}$$

where m_q denotes the mass of the average quark mass. For quark gas the average quark mass can be calculated according to formula [2]

$$m_q=\frac{m_u+m_d+m_s}{3}= \\ \frac{350\text{ MeV}+350\text{ MeV}+550\text{ MeV}}{3}=417\text{ MeV}, \tag{20.34}$$

where m_u, m_d, m_s denotes the mass of the up, down and strange quark respectively. For the calculation of the α_s^q we consider the decays of the baryon resonances. For strong decay of the Σ^0(1385 MeV) resonance:

$$K^- + p \rightarrow \Sigma^0(1385\text{ MeV}) \rightarrow \Lambda + \pi^0$$

width $\Gamma \sim 36$ MeV and lifetime τ_s.

For electromagnetic decay

$$\Sigma^0(1192\text{ MeV}) \rightarrow \Lambda + \gamma,$$

$\tau_e \sim 10^{-19}$ s. Considering that

$$\left(\frac{\alpha_s^q}{\alpha}\right) \sim \left(\frac{\tau_e}{\tau_s}\right)^{\frac{1}{2}} \sim 100, \tag{20.35}$$

one obtains for α_s^q the value

$$\alpha_s^q \sim 1. \tag{20.36}$$

Substituting formula (20.35),) to formula (20.33) one obtains:

$$E_h^q \sim 139\,\mathrm{MeV} \sim m_\pi, \tag{20.37}$$

where m_π denotes the π – meson mass. It occurs that when we attempt to "melt" the nucleon in order to obtain the free quark gas the energy of the *heaton* is equal to the π – meson mass (which consists of two quarks). It is the simple presentation of quark confinement. Moreover it seems that the standard approaches to the melting of the nucleons into quarks through the heating processes in "splashes" of the chunks of the nuclear matter do not promise the success.

QGP AS NON-NEWTONIAN FLUID

We argue that at excitation energy of the order of pion mass the hadronic matter undergoes the phase transition to the non-Newtonian fluid with the relaxation time, τ

$$\tau = \frac{\hbar}{m_q c^2}, \tag{20.38}$$

where $m_q c^2$ is of the order of 400 MeV and the velocity of sound $\upsilon_h \sim c$. The thermal energy quanta in non-Newtonian hadronic fluid is of the order of 140 MeV $\cong m_\pi$. In this model the π – meson can be described as the "phonon" – excitations of the non-Newtonian hadronic fluid. With the excitation energy higher then 140 MeV the boiled fluid evaporates the π – meson copiously. The spectra of the emitted π – mesons can be described by formula [1]

$$d\eta = \frac{N_0 \frac{m}{T}}{K_2\left(\frac{m}{T}\right)} c^{-3} \gamma^5 u^2 \exp\left[-\frac{m\gamma}{T}\right] dV du. \tag{20.39}$$

In formula (40) $\gamma = \left(1 - \frac{u^2}{c^2}\right)^{-\frac{1}{2}}$, T is the temperature of the fluid and dV is the volume element. Function $K_2\left(\frac{m}{T}\right)$ is the modified Bessel function of the second kind.

REFERENCE

[1] M. Kozlowski, J. Marciak-Kozlowska, arXiv 0306 046.

PART III. FIELD EQUATION FOR LIFE SCIENCE

Chapter 21

FIELD INDUCED PROCESSES IN *IN VITRO* CANCER TUMOR

Physicists have long been at the forefront of cancer diagnosis and treatment, having pioneered the use of X rays and radiation therapy. In the contemporary initiative, the US National Cancer Institute the conviction that physicists bring unique conceptual insights that could augment the more traditional approaches to cancer research is very appealing.

In this paper we present the first attempt to consider the tumor cancer as the physical medium with some sort of memory – thermal memory.

The interaction of the short pulses laser, shorter then the relaxation time generates the thermal processes – thermal waves which can propagate through a cancer body.

In this paper we formulate the theoretical framework for the study of the propagation of thermal wave, its velocity and relaxation time.

With the help of our hyperbolic heat transport equation we calculate the temperature field in tumor. As the result we obtain the value for thermal wave, $\upsilon = 104\ \mu$m/s, relaxation time $\tau \sim 10$ s. In contrast to thermal diffusion, in the thermal wave, the thermal energy is well localised, which can be very important to the future therapy of the tumors.

MEMORY OF THE CANCER CELLS

Cancer is pervasive among all organisms in which adult cells proliferate. There is Darwinian explanation of cancer insidiousness which is based on the fact that all life on Earth was originally single-celled.

Each cell had a basic imperative: replicate, replicate, replicate. However, the emergence of multicellular organisms about 550 millions years ago required individual cells to co-operate by subordinating their own selfish genetic agenda to that of the organism as a whole. So when an embryo develops, identical stern cells progressively differentiate into specialized cells that differ from organ to organ.

If a cell does not respond properly to the regulatory signals of the organism it may go reproducing in an uncontrolled way, forming a tumor specific to the organ in which it arises. A key hallmark of cancer is that it can also grow in an organ where it does not belong: for example a prostate cancer cell may grow in a lymph mode.

This spreading and invasion processes is called "metastasis". Metastatic cells may lie dormant for many years in foreign organs evading the body's immune system while retaining their potency. Healthy cells, in contrast, soon die if they are transported beyond their rightful organ.

In some respect, the self centered nature of cancer cells is a reversion to an ancient pre-multicellular lifestyle. Nevertheless cancer cells do co-operate to a certain extent. For example tumors create their own new bloody supply, a phenomenon called "angiogenesis" by co-opting the body's normal wound healing functions. Cancer cells are therefore neither rogue "selfish cells", nor do they display the collective discipline of organism with fully differentiated organs. They fall somewhere in between perhaps resembling an early form of loosely organized cell colonies. In other words the cancer tumor remember the early state of existence, it has a memory which have been erased in healthy cells.

The proliferation of the tumor cells is described by the diffusion processes [1]. The standard diffusion equation is based on the Fourier law in which as we know all memory of the initial state is erased.

Simply speaking diffusion equation has not time reversal symmetry, i.e. if the function $f(x,t)$ is the solution of Fourier equation, $f(x,-t)$ is not.

THE *IN VITRO* STUDY OF THERMAL PROCESSES IN CARCINOMA SPECIMEN

In paper [1] the thermal energy absorption in carcinoma specimen was presented. After the surgery (mastectomy) the breast cancer specimen was isolated. The specimen three centimeters in diameter was irradiated with RF (Radio Frequency) energy.

In the study the following results were obtained. First of all the excised breast tissue showed a preferential heating of the tumor. The fat between tumor strands was surprisingly unaffected. The degree of preferential heating depended on tissue properties tumor shape.

In paper [1] the authors used the Fourier equation for the computer model of the thermal phenomena in the breast cancer. It is well known that the Fourier parabolic equation describes only the slow thermal diffusion processes.

In this paragraph we propose the generalization of the study with the hyperbolic heat transport equation instead of the Fourier equation.

The model equation for the heat transport, with $V = 0$ can be written as [2]

$$\frac{\partial T}{\partial t} = D_T \nabla^2 T - \frac{D_T}{\upsilon_s^2}\frac{\partial T}{\partial t} \tag{21.1}$$

In equation (21.1) D_T is the thermal diffusivity and υ_s is the velocity of the thermal propagation. In the case of Fourier equation $\upsilon_s = \infty$. The solution of (3.1) for 1D is given by

$$T(x,t)=\frac{1}{\upsilon_s}\int dx' T(x',0)\left[\begin{array}{l} e^{-t/2\tau}\frac{1}{t_0}\Theta(t-t_0)+ \\ e^{-t/2\tau}\frac{1}{2\tau}\left\{\begin{array}{l} I_0\left(\frac{\left(t^2-t_0^2\right)^{1/2}}{2\tau}\right)+ \\ \frac{t}{\left(t^2-t_0^2\right)^{1/2}} I_1\left(\frac{\left(t^2-t_0^2\right)^{1/2}}{2\tau}\right) \end{array}\right\}\Theta(t-t_0) \end{array}\right] \tag{21.2}$$

where υ_s is the thermal wave propagation, $t_0=(x-x')/\upsilon_s$, $I_0(z)$, $I_1(z)$ are modified Bessel functions and $\Theta(t-t_0)$ denotes the Heaviside function. We are concerned with the solution t_o (21.1) for a nearly delta function temperature pulse generated by laser irradiation of the medium. The pulse transferred to the specimen has the shape

$$\begin{aligned} \Delta T_0 &= \frac{\beta\rho_E}{c_V\upsilon_s\Delta t} \quad \text{for } 0\le x<\upsilon_s\Delta t \\ \Delta T_0 &= 0 \qquad\qquad \text{for } x\ge \upsilon_s\Delta t \end{aligned} \tag{21.3}$$

In equation (21.3) ρ_E denotes the heating pulse fluence, β is the efficiency of the absorption of energy in the solid, c_V is the heat capacity and Δt is the duration of the pulse. With $t = 0$ temperature profile described by the (21.3) yields

$$\begin{aligned} T(l,t) &= \tfrac{1}{2}\Delta T_0 e^{-t/2\tau}\Theta(t-t_0)\Theta(t_0+\Delta t-t) \\ &+ \tfrac{\Delta t}{4\tau}\Delta T_0 e^{-t/2\tau}\left\{I_0(z)+\frac{t}{2\tau}\frac{1}{z}I_1(z)\right\}\Theta(t-t_0), \end{aligned} \tag{21.4}$$

where $z=\left(t^2-t_0^2\right)^{1/2}/2\tau$ and $t=l/\upsilon_s$. The solution of t_o (21.2) when there are reflecting boundaries is the superposition of the temperature at l from the original temperature and from image heat source at $\pm 2nl$. This solution is

$$T(l,t)=\sum_{i=0}^{\infty}\left[\begin{array}{l} \Delta T_0 e^{-t/2\tau}\Theta(t-t_i)\Theta(t_i+\Delta t-t) \\ +\Delta T_0\frac{\Delta t}{2\tau}e^{-t/2\tau}\left\{I_0(z_i)+\frac{t}{2\tau}\frac{1}{z_i}I_1(z_i)\right\}\Theta(t-t_i) \end{array}\right], \tag{21.5}$$

where $t_i=t_0,\ 3t_0,\ 5t_0,\ t_0=l/\upsilon_0$

Model Calculations

Considering the existing values of the parameters c_V, ρ and k (Table 1) the diffusion coefficient can be calculated [2]

$$D = \frac{k}{c_V \rho} = 8.0 \cdot 10^{-8} \mathrm{m^2 s^{-1}} \tag{21.6}$$

Assuming the known [1] values of the relaxation time for biological materials, τ = 10 s the velocity of the thermal energy propagation can be calculated [2]

$$\upsilon = \sqrt{\frac{D}{\tau}} = 100 \mu\mathrm{m/s} \cdot \tag{21.7}$$

Table 1. Cancer thermal parameters

Tumor	Thermal conductivity W (m°C)	Specific heat kJ/(kg°C)	Density kg/m^3	Thermal diffusion coefficient
Carcinoma	0.28 [1]	3.5 [2]	1000 [2]	$8 \cdot 10^{-8}$ Calculated (this paper)

[1] N. M. Sudarshan et al., *Comput Methods Biomech. Biomed. Engin. 2* (1999) 187; M. Gautherie et al., Biomedicine *22* (1975) 237.

[2] B. Erdmann, *Ann NY Acad. Sci. 858* (1998) 36.

References

[1] V. Ekstand et. Al., *BioMedical Engineering OnLine,* 4:41, (2005) 1 (and references therein).

[2] M. Kozlowski, J. Marciak-Kozlowska, *Thermal processes using attosecond laser pulses,* Springer, USA 2006.

Chapter 22

FIELD INDUCED PROCESSES IN THE HUMAN CORNEA

Basically, laser pulse can interact with biological tissue in five ways: (a) the electromechanical mode, (b) ablation, (c) photothermal (congulative and vaporizing) processes, (d) photochemical (photodynamics) reaction, (e) biostimulation and wound heating.

In this paper we will study in detail the photothermal processes. In the first approximation the cornea is the semiconductor and the electromagnetic field of the laser pulse generates the transport of charge, mass and heat in cornea. These transport phenomena in first approximation can be described as the diffusion processes.

Knowledge on heat transfer in living tissues has been widely used in therapeutic applications. For further studying thermal behavior in biological bodies, many models describing bioheat transfer have been developed [1]. Due to simplicity and validity, the Pennes model is the most commonly used one among them. The applications of this relatively simple bioheat equation include simulations of hyperthermia and cryosurgery, thermal diagnostics. The Pennes bioheat equation describes the thermal behavior based on the classical Fourier's law. As is well known, Fourier's law depicts an infinitely fast propagation of thermal signal, obviously incompatible with physical reality [2]. Thus a modified flux model for the transfer processes with a finite speed wave is suggested and solve the paradox occurred in the classical model. This thermal wave theory introduces a relaxation time that is required for a heat flux vector to respond to the thermal disturbance (that is, temperature gradient) as The relaxation time is approximated as $\tau = \alpha / V^2$. Here, α is the thermal diffusivity, and V denotes the heat propagation velocity in the medium. In homogenous materials such as common metals, the relaxation time ranges from 10^{-8} to 10^{-14} s The heating processes are mostly much longer than this time scale. This is why the phenomenon of the heat wave is difficult to observe in homogenous substances. In reality, the living tissues are highly nonhomogenous, and accumulating enough energy to transfer to the nearest element would take time. The literatures reported the value of τ in biological bodies to be 20–30 s. Mitra et al. found the relaxation time for processed meat is of the order of 15 s. Recently, Roetzel et al. experimentally investigated the relaxation thermal behavior in nonhomogenous materials.

EYE SURGERY WITH ULTRA-SHORT LASER PULSES

One of the earliest and most successful application of short laser pulses is laser eye surgery. To correct near or farsightandness in a person portion of the cornea are cut and reshaped so that the cornea will then correctly focus. In photorefractive kertectomy UV laser light is used to photoablate and reshape of cornea but also ablates the protective endothelial layers on the corneal surface. In a modification of this procedure laser assisted in situ keratomileusis {LASIK}, }, the corneal tissue is first exposed by cutting a flap from the surface layer with a mechanical blade. Then UV light is used to photoablate the corneal tissue and the flap is replaced. The preservation of the surface endothelial layer helps speed recovery from the surgery. While UV laser ablation via linear absorption was a great improvement other mechanical cutting of the corneal tissue, new technique using femtosecond laser ablation offer further improvements in the surgical outcome.

Juhasz et al. developed an optical technique to cut this flap by taking advantage of the nonlinearity of ablation by femtosecond laser. The laser based cutting of the flap is far more precise and is now performed at many clinics and offers significant reduction in side effects [2] The penetration of laser thermal energy in *cornea* strongly depends on the underlaing mechanism. In this paper we describe new mode of cornea heat transport i.e. thermal waves

When exposed to attosecond laser pulses cornea behaves as the multi-layered structure. On the edges of the layers the thermal waves are reflected. Moreover the thermal energy is focused on the front of the thermal wave and can be localized. In the diffusion model the thermal energy is smoothly distributed over all heating region. The localization of the energy in front of the thermal wave enables the control of the heating processes.

MATHEMATICAL FORMULATION

High-energy beams like X-rays and lasers are being increasingly used in a variety of material processing, manufacturing and biomedical applications. Recently developed short pulses lasers such as femtosecond and attosecond lasers, have the additional ability to allow studies of matter on the atomic level

The non-zero value of τ changes the mathematical structure of the transport equation; for parabolic partial differential equations (PDE) such as Fick's law and Fourier's law, are not valid when $\tau \neq 0$. In this case a more general hyperbolic transport equation like the non-Fourier equations are applied to the study of such transport processes [2] According to the constitutive relationship in the non-Fourier processes, heat, mass or charge flux, q, obey the relationship

$$\vec{q}_T(r,t+\tau_T) = -k\nabla T \tag{22.1}$$

and for thermal processes characterized by temperature, T:

$$\vec{q}_n(r,t+\tau_n) = -D\nabla n \tag{22.2}$$

for diffusion of species characterized by density, n. In Equation (1) and Equation (2), k and D denote the thermal conductivity and diffusion coefficient respectively. The terms $\tau_{n,T}$ denotes the relaxation time for density and heat diffusion, respectively. The gradients ∇T and ∇n established in the material at time t result in a flux that occurs at later time, $t + \tau$, due to the insufficient time of response.

The Taylor's series expansion when applied to the Equation (1) and Equation (2) gives

$$\vec{q}_{n,T}(\vec{r},t) + \tau_{n,T}\frac{\partial \vec{q}_{n,T}(\vec{r},t)}{\partial t} + \frac{\partial^2 \vec{q}_{n,T}(\vec{r},t)}{\partial t^2}\frac{\tau_{n,T}^2}{2} = -B_{n,T}\nabla\begin{pmatrix} n(\vec{r},t) \\ T(\vec{r},t)\end{pmatrix} \tag{22.3}$$

where $B_n = D$ and $B_T = k$. In the linearized theory of the transport phenomena the time lag is assumed to be small and the higher order terms in Equation (22.3) can be neglected. By retaining only the first order term in τ, Equation (22.3) can be written as

$$\vec{q}_{n,T}(\vec{r},t) + \tau_{n,T}\frac{\partial \vec{q}_{n,T}(\vec{r},t)}{\partial t} = -B_{n,T}\nabla\begin{pmatrix} n(\vec{r},t) \\ T(\vec{r},t)\end{pmatrix} \tag{22.4}$$

By combining Equation (4) with conservation laws the hyperbolic transport equations for n and T can be obtained [1]

As a simple illustration of the hyperbolic transport processes let us consider the model independent hyperbolic equation - the Heaviside equation

$$\tau\frac{\partial^2 u}{\partial t^2} - c^2\frac{\partial^2 u}{\partial x^2} + \frac{\partial u}{\partial t} = 0 \tag{22.5}$$

where $\tau > 0$, but if we set $\tau = 0$ then Equation (22.5) formally reduces to the parabolic equation. The hyperbolic equations of the type shown by Equation (22.5) are used in physics, biophysics and archaeology We might expect that as $\tau \to 0$ the causal fundamental solution of Equation (5) corresponding to a source point at $P_0(x_0, t_0)$ reduces to the causal fundamental solution of the parabolic equation. Indeed, the causality condition for both problems requires that $u(x,t) = 0$ for $t < t_0$. Furthermore, the causal fundamental solution of Equation (22.5) vanishes outside the forward characteristic sector $|x - x_0| < \hat{c}(t - t_0)$, where $\hat{c} = \frac{c}{\sqrt{\tau}}$. As $\tau \to 0$ the boundary, $\sqrt{\varepsilon}|x - x_0| = c(t - t_0)$, of this sector tends to the line $t = t_0$ which is a characteristic line for the parabolic equation. In the limit, therefore, the fundamental solution of Equation (5) is expected to be non-zero in the region, $t > t_0$ - a result appropriate for the parabolic equation. So, let

$$u(x,t) = \exp\left[-\frac{1}{2\varepsilon}(t - t_0)\right]\upsilon(x,t) \tag{22.6}$$

Substitution Equation (22.5) to Equation(22.6) one `s obtains

$$\tau \frac{\partial^2 \upsilon}{\partial t^2} - c^2 \frac{\partial^2 \upsilon}{\partial x^2} - \frac{1}{4\varepsilon}\upsilon = 0 \tag{22.7}$$

The solution of Equation (22.7) has the form

$$u(x,t) = \frac{1}{\sqrt{4c^2\varepsilon}} \exp\left[-\frac{(t-t_0)}{2\varepsilon}\right] I_0\left[\frac{1}{\sqrt{4c^2\varepsilon}}\sqrt{\frac{c^2}{\varepsilon}(t-t_0)^2 - (x-x_0)^2}\right]$$

$$\text{for } |x-x_0| < \frac{c}{\sqrt{\varepsilon}}(t-t_0) \tag{22.8}$$

$$u(x,t) = 0 \qquad \text{for } |x-x_0| > \frac{c}{\sqrt{\varepsilon}}(t-t_0)$$

As $z \to \infty$ the modified Bessel function, $I_0(z)$, has asymptotic behaviour:

$$I_0(z) \approx \frac{1}{\sqrt{2\pi\, z}} e^{z}, \qquad z \to \infty \tag{22.9}$$

Also, for $t > t_0$

$$\sqrt{\frac{c^2}{\varepsilon}(t-t_0)^2 - (x-x_0)^2} = \frac{c}{\sqrt{\varepsilon}}(t-t_0)\left[1 - \frac{\varepsilon}{2c^2}\frac{(x-x_0)^2}{(t-t_0)} + \ldots\right] \tag{22.10}$$

Thus

$$\left(4c^2\varepsilon\right)^{½}\sqrt{\frac{c^2}{\varepsilon}(t-t_0)^2 - (x-x_0)^2} = \frac{t-t_0}{2\varepsilon} - \frac{(x-x_0)^2}{4c^2(t-t_0)} + \ldots \tag{22.11}$$

As $\tau \to 0$ in Equation (22.8) the argument of I_0 tends to infinity. Using Equations (22.9), (22.10) and (22.11) we obtain in the limit as $\tau \to 0$

$$u(x,t) = \frac{1}{\sqrt{4\pi c^2(t-t_0)}} \exp\left[-\frac{(x-x_0)^2}{4c^2(t-t_0)}\right] \qquad \text{for } \quad t > t_0$$

$$u(x,t) = 0 \qquad \text{for } \quad t < t_0 \tag{22.12}$$

which is identical to the fundamental solution of the parabolic equation.

We conclude that for $\tau \neq 0$ the Heaviside equation (Equation (22.5)) is the generalized diffusion equation. The parameter τ has the dimension of time and is recognized as the relaxation time for the processes described by the Heaviside equation. Considering that in all realistic transport phenomena $\tau \neq 0$ then the Heaviside equation and the more general Proca-Klein–Gordon equation are the master equations for those processes. *Hyperbolic versus parabolic*In the description of the evolution of any physical system it is mandatory to evaluate, as accurately as possible, the order of magnitude of different characteristic time scales because their relationship with the time scale of observation (the time during which we assume our description of the system to be valid) will determine, along with the relevant equations, the evolution pattern. Take a forced damped harmonic oscillator and consider its motion on a time scale much larger than both the damping time and the period of the forced oscillation. Then, what one observes is just a harmonic motion. Had we observed the system on a time scale of the order of (or smaller) than the damping time, the transient regime would have become apparent. This is rather general and of a very relevant interest when dealing with dissipative systems. It is our purpose here, by means of examples and arguments related to a wide class of phenomena, to emphasize the convenience of resorting to hyperbolic theories when dissipative processes, either outside the steady-state regime or when the observation time is of the order or shorter than some characteristic time of the system, are under consideration. Furthermore, as it will be mentioned below, transient phenomena may affect the way in which the system leaves the equilibrium, thereby affecting the future of the system even for time scales much larger than the relaxation time.

Parabolic theories of dissipative phenomena have a long and venerable history, and have proved very useful, especially in the steady-state regime. They do, however, exhibit some undesirable features, such as acausality [1]. This has prompted the formulation of hyperbolic theories of dissipation to eradicate these undesirable features. But, this was achieved at the price of extending the set of field variables by including the dissipative fluxes (heat current, non-equilibrium stresses and so on) on the same footing as the classical ones (energy densities, equilibrium pressures, etc), thereby giving rise to a set of more physically satisfactory (as they much better conform with experiments) but involved theories from the mathematical point of view. These theories have the additional advantage of being backed by statistical fluctuation theory, kinetic theory of gases (Grad's 13-moment approximation}

A key quantity in these theories is the τ of the corresponding dissipative process. This positive-definite quantity has a distinct physical meaning; namely the time taken by the system to return spontaneously to the steady state (whether of thermodynamic equilibrium or not) after it has been suddenly removed from it. It is, however, connected to the mean collision time, t_c, of the particles responsible for the dissipative process. It is therefore appropriate to interpret ε as the time taken by the corresponding dissipative flow to relax to its steady value. Thus, it is well known that the classical Fourier law for heat current [2],

$$\vec{q} = -\kappa \nabla T \tag{2,13}$$

with q being the heat conductivity of the fluid, leads to a parabolic equation for temperature (diffusion equation):

$$\frac{\partial T}{\partial t} = \chi \, \nabla^2 T, \qquad \chi = \frac{\kappa}{\rho \, c_V} = \frac{\kappa}{C_V}; \qquad C_V = \rho \, c_V . \tag{22.14}$$

where χ, ρ and c_V are diffusivity, density and specific heat at constant volume, respectively, which does not forecast propagation of perturbations along characteristic causal light-cones; that is to say, perturbations that propagate with infinite speed. This non-causal behaviour is easily visualized by taking a look at the thermal conduction in an infinite one-dimensional medium. Assuming that the temperature of the line is zero for $t < 0$, and putting a heat source at $x = x_0$ when $t = 0$, the temperature profile for $t > 0$ is given by

$$T \sim \frac{1}{\sqrt{t}} \exp\left[-\frac{(x - x_0)^2}{t} \right] \tag{22.15}$$

Equation (15) thus implies that $t = 0 \Rightarrow T = \delta(x - x_0)$, and for $t = t_1 > 0 \Rightarrow T \neq 0$. In other words, the presence of a heat source at x_0 is instantaneously felt by all observers on the line, no matter how far away from x_0 they may happen to be. The origin of this behaviour can be traced to the parabolic character of Fourier's law, which implies that the heat flow starts (vanishes) simultaneously with the appearance (disappearance) of a temperature gradient. Although τ is very small for phonon-electron, and phonon-phonon interaction at room temperature (10^{-11} and 10^{-13} s, respectively), neglecting it is the source of difficulties and in some cases a bad approximation; for example in super-fluid He and degenerate stars where thermal conduction is dominated by electrons [1].

In order to overcome this problem Cattaneo and (independently) Vernotte, by using the relaxation time approximation to Boltzmann equation for a simple gas, derived a generalization of Fourier's law [1]:

$$\tau \frac{\partial \vec{q}}{\partial t} + \vec{q} = -\kappa \, \nabla T \tag{22.16}$$

This expression (known as the Cattaneo-Vernotte's equation) leads to a hyperbolic equation for the temperature which describes the propagation of thermal signals with a finite speed:

$$\upsilon = \sqrt{\frac{\chi}{\tau}} \tag{22.17}$$

This diverges only if the unphysical assumption of setting □ to zero is made. It is worth mentioning that a simple random walk analysis of transport processes naturally leads to a hyperbolic equation, not to the diffusion equation [2]. Again, the latter is obtained only if one neglects the second derivative term.

It is instructive to write (16) in the equivalent integral form:

$$\vec{q} = -\frac{\chi}{\tau}\int_{-\infty}^{t}\exp\left[-\frac{(t-t')}{\tau}\right]\cdot\nabla T(x',t')dt' \tag{22.18}$$

which is a particular case of the more general expression

$$\vec{q} = -\int_{-\infty}^{t}K(t-t')\nabla T(x',t')dt' \tag{22.19}$$

The physical meaning of the kernel $K(t - t')$ becomes obvious by observing that for

$$K = \kappa\delta(t-t') \rightarrow \vec{q} = -\kappa\nabla T \text{ (Fourier)} \tag{22.20}$$

and for

$$K = \text{constant} \rightarrow \frac{\partial^2 T}{\partial t^2} = \chi\nabla^2 T \text{ (wave motion)} \tag{22.21}$$

with K describing the thermal memory of the material by assigning different weights to temperature gradients at different moments in the past. The Fourier law corresponds to a zero-memory material (the only relevant temperature gradient is the "last" one; that is the one simultaneous with the appearance of q). By contrast, the infinite memory case (with K = constant) leads to an undamped wave. Somewhere in the middle is the Cattaneo-Vernotte equation, for which all temperature gradients contribute to q, but their relevance goes down as we move to the past. As the third case, "intermediate memory" will be considered:

$$K(t-t') = \frac{K_3}{\tau}\exp\left[-\frac{(t-t')}{\tau}\right] \tag{22.22}$$

By combining Equation (22.22) and Equation (22.18) we obtain

$$c_V\frac{\partial^2 T}{\partial t^2} + \frac{c_V}{\tau}\frac{\partial T}{\partial t} = \frac{K_3}{\rho\tau}\nabla^2 T \tag{22.23}$$

and

$$K_3 = D_T c_V \rho \tag{22.24}$$

Thus, finally

$$\frac{\partial^2 T}{\partial t^2} + \frac{1}{\tau}\frac{\partial T}{\partial t} = \frac{D_T}{\tau}\nabla^2 T \tag{22.25}$$

$D_T = v^2\tau$

where v is the velocity of heat propagation. From these comments it should be clear that different classes of dissipative systems may be described by different kernels. Obviously, when studying transient regimes; that is, the evolution from a steady-state situation to a new one, τ cannot be neglected. In fact, leaving aside that parabolic theories are necessarily non-causal, it is obvious that whenever the time scale of the problem under consideration becomes of the order of (or smaller) than τ , the latter cannot be ignored. It is common sense what is at stake here: neglecting τ in this situation is to disregarding the whole problem under consideration. According to a basic assumption underlying the disposal of hyperbolic dissipative theories, dissipative processes with values of τ comparable to the characteristic time of the system are out of the hydrodynamic regime.

For medical and biological sciences the parabolic heat diffusion equation was developed and solved by Pennes In the paper [2]we have shown the general method for introducing the generalised hyperbolic Pennes equation.

REFERENCES

[1] Kozlowski M., Marciak – Kozlowska J., From femto-to attoscience and beyond, Nova Science Publishers, 2009,USA

[2] Kozłowski M., Marciak-Kozłowska J. ar xiv.1108.0898

Part IV. Consciousness Field

Overview of the Research

This part of the book discusses historical and sociological aspects of the relationship between paraphysics and parapsychology. The treatment is somewhat superficial and focuses mainly on the ESP, but it will suffice for present purposes. More detailed accounts of the history of the ESPhave been given by Alan Gauld and Renee Haynes).

The key figures in the ESP have always included both psychologists and physicists, but their relative prominence has changed considerably over the 126 years since its founding. This can be seen by examining the list of past Presidents, which is fairly representative of both the general membership of the ESP and the field as a whole. Of course, the subject also involves other scientific disciplines — including biology, for which psi processes may be equally fundamental. However, psychology and physics might be regarded as the two poles in the usual 'reductionist' classification of the sciences (discussed later).

Unfortunately, the study of the paranormal has not gained academic acceptability within physics departments (or any other 'hard' science departments) in the same way that it has within psychology departments. Indeed the study of paraphysics is still academically taboo: the only professional physicist who works on the subject in a UK university is Professor Brian Josephson at Cambridge, and even he does not focus exclusively on the paranormal. Furthermore, no PhDs have been obtained in the subject in UK physics departments—not even in Cambridge. On the other hand, the situation in paraphysics today is really no worse than it was in parapsychology 20 years ago, when Edinburgh was the only active department. In principle, therefore, another 20 years might see an equally dramatic proliferation of paraphysics groups. One just needs someone to do for paraphysics what Robert Morris did for parapsychology.

Of course, many physicists are *interested* in the subject, at least to the extent of publishing articles about it. They number several dozen in the UK and around a hundred worldwide. The work of many of them will be cited in this Chapter. Even so, they represent only a tiny fraction of the total physics community and their paraphysical work is usually conducted in their spare time. The few professional physicists who are paid to work in the subject are not generally university-based. In any case, physicists who speculate in this area—whether or not they have a university affiliation—are liable to be regarded with suspicion by their peers.

On the other hand, there are some positive signs. Despite the situation in university departments, there is evidence that physicists may be more open to the occurrence of psi than psychologists: a survey of US and Canadian academics some decades ago found that 55% of physical scientists thought psi was possible, compared with 34% of psychologists (McClenon, 1982). Another welcome development is that, besides the specialist parapsychological journals, there are now a number of more general science journals which include articles about paraphysics. These include the *Journal of Consciousness Studies* (which in 2003 and 2005 devoted entire issues to parapsychology) and the *Journal of Scientific Exploration.* There are also two electronic journals covering the subject: Beichler's *Yggdrasil: The Journal of Paraphysics* and Lian Sidorov's *Journal of Non-Locality and Remote Mental Interactions.*

The aversion of physicist to pschical research Even in the early days of psychical research, physicists who took the paranormal seriously and tried to link it to physics attracted hostility from their mainstream colleagues. Crookes's publications on the subject were much ridiculed, even though he was a most distinguished physicist and later became President of the Royal Society. His observations of materializations during experiments with Florence Cook were once even attributed to poisoning by thallium — the element he had discovered! Lodge received a lot of criticism for publishing a paper on telepathy in *Nature*, and Barrett's attempts to set up a committee of the British Association to investigate the subject were rejected outright.

Many physicists remain antagonistic towards parapsychology in the present age. In 1979- a symposium on psi and physics was hosted by the American Association for the Advancement of Science. This attracted intense opposition from John Wheeler, who attempted to eject the Para- psychological Association from the AAAS with the battle-cry "Drive the pseudos out of science ... . Where there's smoke, there's smoke" (Wheeler). More recently, Gerard 't Hooft, who won the Nobel prize for physics in 1999 and runs an anti-parapsychology website, has stated ('t Hooft,):-

Modern physics seems to offer leeway to the paranormal. As a theoretical physicist, I must assert most emphatically that this leeway is only apparent. There is absolutely no way one can explain the paranormal in this fashion.

The aversion of some physicists to parapsychology was vividly illustrated some years ago by a furore involving the ESP's Nobel Laureate, Brian Josephson. In September 2001 the Post Office issued a set of stamps commemorating the centenary of the Nobel Prize, with one for each of the six subject areas in which the prizes are awarded. This was accompanied by the publication of a brochure in which various UK laureates— including Josephson — were asked to provide a brief commentary on the area involved in their discovery, together with a look to the future. Josephson used this oportunity to suggest that quantum theory may one day lead to an understanding of telepathy and the paranormal:-

Quantum theory is now being combined with theories of information and computation, These developments may lead to an explanation of processes still not understood within conventional science, such as telepathy, an area where Britain is at the forefront of research. This provoked some hostile responses. An article in the *Observer* on 30

September contained an onslaught from the physicist David Deutsch, who dismissed Josephson's claims outright:-

Telepathy simply does not exist. ... The evidence for its existence is appalling ... . The Royal Mail has let itself be hoodwinked into supporting ideas that are complete nonsense.

Although Deutsch is a renowned quantum physicist, his brash dismissal of the evidence for telepathy makes one wonder to what extent he has studied this or indeed read much about the subject at all. Other sceptics soon joined the fray. In the same *Observer* article the previous year's physics Nobel Laureate, Herbert Kroemer, declared *"Few of us believe telepathy exists, nor do we think physics can explain it"*. Then, on Radio 4's *Today* programme, Josephson did battle with James Randi, who (though not a physicist himself) declared that *"trying to explain ESP with quantum mechanics is the refuge of scoundrels"*.

Why does such hostility arise? One obvious factor is doubts about the strength of the evidence and the fact that—according to an influential paper by Irwin Langmuir (1989)—parapsychology shares some features of pathological science. He lists these as follows: (1) the maximum effect is barely detectable; (2) many measurements are necessary because of the low statistical significance of the results; (3) fantastic theories are constructed contrary to experience; (4) criticisms are met by *ad hoc* excuses; and (5) the ratio of supporters to critics rises to nearly 50% and then gradually falls to zero.

There is a clear contradiction between science and most supernatural phenomena... . The entire edifice of physics would have to be reconstructed from the ground up if it had to embrace psi phenomena. In this context, of course, one must distinguish between what is *compatible* with physics and what is *explicable* by it. Many psi phenomena may be irrelevant to physics, and even telepathy might be if one adopts a dualist philosophy in which mind-mind interactions do not reduce to brain-brain interactions. The problem is that many psychic phenomena *do* involve an interaction with the physical world, and furthermore appear to violate such cherished notions as causality.

The second reason why many physicists are antagonistic towards psi is the implied threat to reductionism. There are many branches of science (e.g. psychology, neurophysiology, biology, chemistry, physics) and for most purposes each of these may be regarded as self-contained, with its own language and conceptual framework. However, according to reductionism, they are logically interdependent and form a hierarchy in which the fundamental concepts and laws at each level can be explained in terms of those pertaining at the lower one.

Of course, many links in the reductionist chain remain controversial. For example, many biologists do not believe that all of biology can be reduced to DNA, and many psychologists do not accept that all mental processes can be explained in terms of neuronal activity. Nevertheless, it is clear that the reductionist outlook is very influential from a sociology of science perspective. In fact, some forms of psi would not necessarily be incompatible with reductionism.

The third reason for physicists' antipathy to psi is that many psychic phenomena involve *consciousness,* and physicists have long been uncomfortable with attempts to incorporate even normal aspects of consciousness (let alone paranormal ones) into physics. This is because the contents of consciousness are intrinsically private, whereas physics deals with what is in the public domain. (Consciousness may still be studied scientifically from the standpoint of social anthropology but that is a separate issue.) Brian Pippard, for example, even though he is open to the possibility of psi, has argued that consciousness will be forever outside the domain of physics

If the existence of these phenomena is doubtful, it is because the evidence is scanty and often of dubious provenance, it is not because they cannot be invoked in physical terms. They involve, after all, a class of system beyond the scope of physical theory—that is to say, conscious human beings.

Certainly physics in its *classical* mechanistic form cannot incorporate consciousness. This was appreciated more than a century ago by William James who stressed the incompatibility between the localized features of mechanism and the unity of conscious experience. Although the classical picture of physics has now been replaced by a more holistic one, and there are some indications that the new physics *can* include consciousness, we will see that this is controversial.

It is not only physicists who are uncomfortable with consciousness. Even some psychologists have been keen to banish any reference to it. Nearly a century ago, the behaviourist John Watson) declared:-

The time seems to have come when psychology must discard all reference to consciousness; when it need no longer delude itself into thinking that it is making mental states the object of observation.

Indeed, since parapsychology was itself born in the behaviourist- dominated period, it might be argued that it was strongly influenced by it. Although attempts by behaviourists to extend mechanism to the mind have now been rejected, a mechanistic outlook still persists among many physicists and this probably contributes to their discomfort with consciousness.

Most physicists would not share this extreme view, but they would still argue that the focus of science should be the objective world, with the subjective element being banished as much as possible (i.e. it should be concerned with the 3rd person rather than the 1st person account of the world). Fortunately, there are now a growing number of physicists interested in the general area of consciousness studies (Hameroff), though most remain sceptical about psi.

*The aversion of psychical researches to physics*It is not only physicists who oppose attempts to link psi and physics. Several prominent psychical researchers are equally uncomfortable with the idea. To quote the late John Beloff (1988):-

The attempt to reconcile physics and parapsychology is misguided. Asking for an explanation of the mind-matter interaction could only lead to an endless and profitless regress.

In fact, in pointing out the alternatives to the paraphysical approach, Beloff goes even further and suggests that psi may be completely anarchic, in the sense that it obeys no laws at all. This would exclude it from the domain of science altogether. Of course, one could always go half-way, accepting that psychic interactions conform to laws which can be studied scientifically but denying that they are part of physics. In particular, this would be compatible with Beloff's dualist view, in which mental and physical phenomena are simply disconnected. This is fine if minds only observe the physical world *passively*. However, some psychical phenomena seem to require minds to influence it *actively* and the laws that govern this interaction must then surely involve physics.

Nevertheless, many psychical researchers share Beloff's scepticism. Some indication of the source of this prejudice comes from Carroll Nash (1986):-

In the sense of being independent of space, time and physical causality, psi is non-physical. Physical causality presumes transmission of energy over time and space between the interacting bodies ... psi's apparent independence of physical causality suggests that, for it, cause and effect may be simultaneous. That psi is not a physical force in the classical sense is indicated by the failure of metal chambers and Faraday cages to prevent its occurrence.

J. B. Rhine was sceptical of a physical theory of ESP for similar reasons. In fact, the claim that psi has been demonstrated to be space-independent and the significance of the Faraday cage experiments are both debatable.

Another cause of antipathy may be the misapprehension that para- physicists wish to describe *all* aspects of mental experience in terms of physics, thereby embracing reductionism. Thus physicists are uncomfortable with paraphysics because it threatens to destroy reductionism, while psychical researchers are uncomfortable with it because it threatens to support reductionism! However, even in the context of 'normal' processes, few people would claim that *every* aspect of mental experience can be reduced to physics. For example, 'secondary' qualities (colour, taste, emotional assocations, etc.) may always transcend it. So it may only make sense to try to extend physics to incorporate those 'primary' features of psychic experience which involve the sensorial contents themselves (e.g. the geometrical features of an apparition rather than its emotional impact).

Finally, of course, it should be stressed that many parapsychologists are not so much antipathetic to physics as uninterested in it. For example, the only physical aspects of psi covered in the most recent edition of Harvey Irwin's classic textbook (Irwin & Watt) are those concerning its mediation.

Reasons for connecting EPS and physics Science assumes that the world is governed by natural laws, and psychical research will only become acceptable to the rest of the scientific community if psychic phenomena are also subject to such laws. The purpose of psychical research should therefore be to demonstrate that natural law can be extended to include psi and not to throw the ball back into the court of the 'supernatural'. If psi turns out to be anarchic, as suggested by Beloff, this aim may be forlorn, but one should at least give it a try. Also chaos theory and non-linear dynamics have taught us that what appears anarchic at one level may turn out to have a discernible pattern at another level.

Now an essential feature of any branch of science is that it must involve some theory to explain the observations, so if psychical research is to qualify one needs a theory for psi. This is why understanding its properties is more important than just accumulating statistical proof of its existence. More recently, Henry Margenau (1985) urged:-

No amount of empirical evidence, no mere collection of facts, will convince all scientists of the veracity and the significance of your reports. You must provide some sort of model: you must advance bold constructs ... in terms of which ESP can be theoretically understood.

There are several historical precedents for this. For example, Alfred Wegener's idea of continental drift was not accepted for several decades because there was no theory to explain it. Although it is not inevitable that a theory for psi has to come from physics (rather than biology, say), it would seem most natural to use the model of the world which already exists and has proved so successful. Also, we have seen that most scientists adopt a reductionist view, so — regardless of whether this is correct—it seems unlikely from a sociological perspective that psi will ever be accepted by mainstream science until it is founded on a theory which connects with physics. Certainly physicists themselves will not accept psi until this happens

Physicists who have retained some humility in the face of nature's mysteries are interested in psi because it implies that we have completely overlooked fundamental properties of space, time, energy and information. Specifically, psi suggests that the conventional boundaries of space and time can be transcended by the ephemeral concept of the 'mind'.

Indeed one of the reasons physicists figured so prominently among the early membership of the ESP was that they saw in psychic phenomena evidence for some new type of physics (Noakes). For the history of physics is full of the inexplicable becoming explicable and studying anomalous effects nearly always leads to useful insights. Thus new phenomena should be welcomed by physicists, even if they are not at first explicable theoretically. For example, it was only several years after its discovery that superconductivity could be explained. Nevertheless, history shows that phenomena which occur only rarely are often received sceptically at first. A good example of this is ball lightning, which was studied by Lord Rayleigh in the 1890s but not acknowledged to be a real phenomenon until the 1960s. On the other hand, it must also be cautioned that new phenomena do sometimes turn out to be spurious (e.g. N-rays).

Now there can be no doubting the success of physics within its own terms. Particularly impressive has been its progressive unification of the different forces of nature, as illustrated in Figure 2. Indeed many people have proclaimed that the end of physics is in sight, in the sense that our knowledge of the fundamental laws and principles governing the Universe is nearly complete. They argue that we are on the verge of obtaining a 'Theory of Everything' (TOE). This description may seem pretentious, because one is really only purporting to have a final theory of *physics,* but we have seen that this may indeed extend to 'everything' if one adopts a reductionist view. On the other hand, physics also seemed close to a complete theory in the 1890s, before the revolutions of relativity theory and quantum mechanics overthrew the classical paradigm, so one should be wary of this claim.

One feature of the Universe which would seem to refute the claim that physics is close to a TOE is the existence of consciousness. We have seen that many physicists regard this as being outside the domain of physics. However, this attitude is not universal and other physicists are equally uncomfortable with attempts to formulate a TOE without any reference to this. Thus Roger Penrose anticipates that "*we need a revolution in physics on the scale of quantum theory and relativity before we can understand mind*", while the linguist Noam Chomsky asserts that "physics must expand to explain mental experiences". It is certainly conceivable that some future paradigm of physics will make an explicit link with mind. We cannot be sure that such a paradigm would accommodate paranormal phenomena — certainly neither Penrose nor Chomsky would advocate this—but one cannot exclude this possibility. If it does, it remains to be seen whether psi involves some new form of 'force', which might eventually be unified with the other forces.

Perhaps the most important reason for wanting to incorporate psi into physics is that many people claim that recent developments in physics already make this possible. The fact that the physical world has turned out to be much weirder than common sense would suggest has led some people to argue that there might well be room for the sort of phenomena studied by parapsychology. To quote Arthur Koestler

The unthinkable phenomena of extra-sensory perception appear somewhat less preposterous in the light of the unthinkable propositions of modern physics. Indeed, Part 2 of this paper will review the many attempts to explain psi in terms of physics explicitly. The general conclusion is that physics is still not weird enough *to accommodate psi, but that some of these attempts may well be relevant.*

Despite these arguments, antipathy to paraphysics is clearly still strong. We therefore face a dilemma. The prime challenge of psychical research is that it needs to link mind and matter. Yet paraphysicists— the people who try to provide this link—find themselves shunned not only by mainstream physicists (who regard the paranormal with scepticism) but even by parapsychologists (who are generally uninterested in such questions).

Is mind fundamental or incidental? Since the Enlightenment, the prevailing scientific view has been that the Universe—and everything within it—behaves like a giant machine, completely oblivious to whether consciousness is present. Recent advances in brain research and artificial intelligence suggest that even the mind may be a machine, and this may be part of the prejudice that psi is impossible. For if free will is an illusion, how can one wilfully influence a physical system? In recent decades, however, there has been a reversal in this view and various arguments now suggest that mind may be a fundamental feature of the Universe rather than an incidental one. This may not directly help to explain psychic phenomena but it does perhaps break down the prejudice that they cannot be real.

One context in which mind has crept into physics is through the *Anthropic Principle.* In its weak form, this just points out that the existence of observers imposes a selection effect on when and where we exist in the Universe. In its (more controversial) strong form, it claims that there are unexplained coincidences involving the physical constants (e.g. the dimensionless numbers which describe the strengths of the four forces) which are required in order that conscious observers can arise. One possible interpretation of the Strong Anthropic Principle is that the Universe is just one member of a huge ensemble of universes, called the 'multiverse' in which the constants vary. In this case, we necessarily inhabit one of the small fraction of anthropically-tuned universes. (Indeed, one multiverse scenario involves the same higher-dimensional theory of physics which I will invoke later to explain psi.) In any case, this suggests that mind may not be entirely irrelevant to the functioning of the Universe.

Another context in which mind may appear in physics involves quantum theory This shows that, on a microscopic scale, matter does not behave like a machine at all. Particles—instead of being localized like billiard balls—are described by a wave-function, and this introduces a new level of randomness and acausality into the world. Quantum theory may also impinge on the mystery of consciousness. Studies of quantum phenomena convinced Louis de Broglie that "*the structure of the material Universe has something in common with the laws that govern the workings of the human mind*". Although opposed to psychical research, Wheeler has inferred from quantum theory that "*mind and Universe are complementary*", while Bernard d'Espagnat claims:

The doctrine that the world is made up of objects whose existence is independent of human consciousness turns out to be in conflict with quantum mechanics and with facts established by experiments. This impression arises because in quantum theory the wave-function which describes any physical system evolves smoothly in accordance - with Schrodinger's equation until a measurement is made. At this point the wave-function is said to 'collapse' to a state which corresponds to a possible result of the measurement. There is some controversy as to what causes this collapse, or what it means, but some people have proposed that consciousness is involved. This is a crucial part of most attempts to relate quantum theory to ESP. Other people assume that quantum-mechanical weirdness does not persist on macroscopic scales and simply cancels out for biological systems. Thus John Hopfield (1990) claims:-

Contrary to the expectations of a long history of ill-prepared physicists approaching biology, there is absolutely no indication that quantum mechanics plays a significant role in biology.

And Murray Gell-Mann asserts that "no vital forces are needed for biology or self-awareness". There is thus the cherished hope that mechanism will survive at the level of the brain itself. However, recent advances in neuroscience may be incompatible with this hope: Stuart Hameroff, and Hameroff and Penrose (1996), propose that quantum effects may occur via microtubules, while Jeffrey Satinover argues that large-scale quantum effects may be captured and amplified by the brain, so that it no longer behaves deterministically. The connection between quantum physics and neuroscience has been recently reviewed in an important paper by Jeffrey Schwartz and colleagues The question of whether quantum effects can explain psi is addressed later.

Can ESP and physics connect? As stressed by David Rousseau there are many ways of classifying psychic phenomena. First, it is important to distinguish between *psychic* and *anomalous* phenomena. Although some parapsychologists prefer the latter term, this is confusing because physicists often discuss anomalies which have nothing to do with parapsychology (e.g. cold fusion). This issue has been discussed by Beichler, who argues against the label 'anomalous'. The usual semantic convention is to regard psychic phenomena as the subset of anomalous phenomena associated with psyche or life, although both these terms are somewhat vague.

- The first class comprises those *alleged* psychic phenomena which are delusional, in the sense that they result from the mind's innate tendency to see patterns in random data (Blackmore & Troscianko It is impossible to give examples of this class without causing offence, but I have tentatively listed pyramid power, the Bermuda triangle and psychic surgery. This selection reflects my own bias (and may be wrong) but I think everybody would agree that *some* psi phenomena are spurious, even if—due to our different 'boggle' thresholds—we disagree on which ones. Were it not for this disagreement, one could dismiss This table classifies (purported) psychic phenomena into four categories. Class 1 are delusional. Class 2 are real but probably explicable by current physics. Class 3 are inexplicable by current physics but apparently involve an interaction with the physical world. Class 4 are purely mental and may have no relevance to physics. Phenomena lie on the boundary of neighbouring classes either because their classification is uncertain or because they may involve a combination of effects. The 2/3 and 3/4 boundaries are expected to evolve with time, as indicated by the arrows. The phenomena inside the double-border are the ones most relevant to paraphysics. All the entries (especially in Class 1) are tentative and reflect my personal bias. The list of phenomena is illustrative and not intended to be complete.
- The second class comprises phenomena that are real but have a simple explanation within the current physical paradigm, despite some people's attempts to endow them with paranormality. I would tentatively include Kirlian photography (a corona discharge effect) and fire-walking (a thermal conduction effect) in this class. Some electronic voice phenomena (EVP) may result from the misinterpretation of normal terrestrial radio transmissions or other indistinct sounds, though advocates would claim not all of them. There is an explicit connection with physics here—indeed any psychical investigator needs to be sufficiently familiar with physics to recognize Class-2 phenomena when they arise. Since the distinction between psychic and anomalous phenomena is somewhat blurred—the connection with psyche is not

always clear-cut and sometimes amounts to no more than the *belief* that mind is involved — a number of anomalous phenomena might also be included in this class.

Some phenomena may be regarded as being on the Class 1/2 boundary because there is uncertainty as to their status. For example, the Loch Ness monster and similar exotic creatures may not exist at all but, even if they do, they presumably have a standard zoological explanation. Likewise, even if some UFO sightings are explained by extraterrestrial visitations, this is more relevant to astronomy than parapsychology. (If these phenomena do not occur on a physical level at all, but are akin to apparitions, they might still be relevant to psi; in this case, they would need to be reclassified, but I am taking the simplest interpretation here.) If accounts of spontaneous human combustion are vindicated, this may have a purely biochemical explanation. If crop circles are explicable by some combination of hoaxes and meteorological effects, they also would be included in this class.

- The third class consists of phenomena which are inexplicable by current physics — and whose reality is therefore controversial — but which nevertheless seem to involve an interaction with the physical world. One would clearly include psychokinesis (PK) in this class and one might also include clairvoyance and precognition if perception of the physical world is assumed to require some signalling mechanism. If minds are generated by brains, as assumed by reductionists, then telepathy would also be included. The emphasis on *current* physics in this definition is, of course, crucial, since one might hope that the boundary between Class-2 and Class-3 would gradually shift as physics advances (so that 'paranormal' phenomena become 'normal').

  Even within the context of the physical paradigm which prevails at a particular time, people will disagree on where the 2/3 boundary comes. An extreme sceptic would regard all psychic phenomena as Class-1 or Class-2, and even people more favourably disposed to the paranormal might like to relegate some of them to Class-2. For example, some people attribute dowsing to electromagnetic effects (in which case it is Class-2), whereas others attribute it to clairvoyance (in which case it is Class-3). Some phenomena could be associated with both Class-2 and Class-3 effects. For example, poltergeist effects may involve a combination of natural factors (like geomagnetic or seismic activity) and recurrent spontaneous psychokinesis (RSPK). Likewise psychic healing (as opposed to psychic surgery, which has a more dubious provenance) and some forms of complementary medicine may involve a combination of some unexplained physical interaction (Class-3) and psychosomatic effects which (from a reductionist standpoint) would be regarded as Class-2. I also include Instrumental Transcommunication (ITC) — a broader term than EVP—on the 2/3 boundary Finally there are a wide range of spiritualistic effects (materializations, apports, spirit photographs, etc.) and magical phenomena (Roney-Dougal), which clearly involve physical manifestations if real.

- The fourth class consists of phenomena which are purely mental, in the sense that they may involve no *direct* interaction with the physical world at all—except perhaps via the brain if one takes the view that all mental experiences are generated by the brain. The existence of this category is not intended to preclude a reductionist view, although Edward Kelly argue that it may. Class fourth includes 'rogue' phenomena, such as hypnosis, hallucinations and multiple-personality manifestations, in this class. Although such states may not be psychic *per se,* they have all come under the scrutiny of psychical researchers at various times because they are sometimes *associated* with psi. Indeed, the whole domain of transpersonal psychology might be included in this class, since there is clearly an overlap between psychic, religious and mystical experiences). Many psychic phenomena—for example, apparitions, death-bed visions 1(DBVs), out-of-body experiences (OBEs), near-death experiences (NDEs), past-life memories and mediumistic phenomena—are placed on the ¾ boundary rather than in Class-4 itself, because it is unclear whether or not they involve an interaction with the physical world (the defining 3 feature of Class-3). The issue here is not the validity of the reports — 3 nobody doubts that the *experiences* are genuine—but their *interpretation.* Is one really out of one's body in an OBE and is one really remembering a past life in hypnotic regression? This relates to whether the experiences on the 3/4 boundary are *veridical.* The problem is that the distinction between Class-3 and Class-4 is fuzzy, since we do not know for sure what is entailed in the terms 'physical world' and 'purely mental'. While it is rclear that there is no room for the contents of mind in the classical world-view, I will argue later that most psi experiences require the existence of some form of 'space', in which case one might hope that the domain of physics could eventually be extended to incorporate this.. This might be regarded as corresponding to a change in the d Cartesian boundary between matter and mind (Beichler, 2006). It should now be clear that the psi phenomena which are relevant to paraphysics are the ones classed as 2/3, 3 and ¾. The crucial question is how far the 2/3 boundary will eventually penetrate into what is currently regarded as Class-3. However, an important caveat should be made here. For if there is a paradigm change, one might anticipate the 2/3 boundary undergoing a single large shift rather than a sequence of small ones. The change involved could then be so radical that the nature of physics might itself change. I would therefore advocate introducing a new term, 'hyper- physics', which is more general than paraphysics since it does not only have implications for psi. In this case, one might want to regard the final limit of the 2/3 boundary as the transition between physics and hyperphysics. Another interesting question is how far the 3/4 boundary will eventually penetrate into what is currently regarded as Class-4. This relates to the eventual status of reductionism; if all mental experiences can ultimately be reduced to physics (or at least hyperphysics), then Class-4 would disappear altogether, although still subjectively defined.

PARADIGM SHIFT AND THE HISTORY OF PHYSICS

The history of science shows that the prevailing model of physical reality regularly undergoes paradigm shifts (Kuhn). The paradigm determines the sort of picture one has of the world, the type of questions one asks about it and the experiments one performs. Much scientific progress is made within the context of a particular paradigm, but eventually anomalies arise and these result in a crisis which ultimately leads to the adoption of a new one. During the crisis, a variety of new theories will be advanced. The upholders of the old paradigm will try to resist these but eventually they die off and the new paradigm takes hold.

- The first paradigm was the classical *Newtonian* one, in which the physical world is regarded as a 3-dimensional continuum, with solid objects moving according to Newton's laws of dynamics. Time is absolute, in the sense that it flows at the same rate for everyone, and there is also an absolute space associated with inertial (non-accelerating) frames. Objects attract each other through the force of gravity, although the paradigm does not explain *why* that force exists.
- The next paradigm, *atomic theory,* arose from developments in statistical physics and thermodynamics. These showed how the interactions of billions of atoms lead naturally to the observed macroscopic laws and how the structure of the atoms themselves provides an understanding of chemistry. The new paradigm also contained the laws of electricity and magnetism. In particular, it showed that light consists of electromagnetic waves travelling through an 'ether', which naturally was identified with Newton's absolute space.
- The advent of the next paradigm, *special relativity*, demolished the idea of the ether and showed that space and time are not absolute but part of a spacetime continuum (called Minkowski space). Thus a consistent picture of how different observers perceive the world requires that it be 4-dimensional rather than 3-dimensional, the fourth dimension being time, and material objects corresponding to worldlines in spacetime.
- The next transformation came with *general relativity*, which showed that spacetime is curved in the presence of matter, like a surface in a higher-dimensional space. This explains the origin of gravity geometrically. It gives different predictions from Newton's theory of gravity but the differences are only large for a strong gravitational field (e.g. for a black hole). General relativity also forms the basis of cosmology, the branch of physics concerned with the structure of the Universe in the large.
- Paralleling these developments in macroscopic physics was the paradigm shift associated with *quantum theory.* This showed that microscopic objects can simultaneously behave like waves and particles. Measurements always interfere with systems in some way and this leads to the Uncertainty Principle. In particular, a particle cannot simultaneously be ascribed a position and velocity, which means that the concept of a worldline (underlying the spacetime description of relativity) can only be an approximation.
- The *Kaluza-Klein* proposal arose out of attempts to give a geometrical explanation of electromagnetic interactions, analogous to the geometrical explanation of gravitation provided by general relativity. Kaluza-Klein theory suggests that the Universe is 5-dimensional; the fifth dimension is wrapped up so small that it cannot be observed directly but its existence neatly explains the laws of electromagnetism. Strictly speaking, this does not yet qualify as a paradigm (since it is not universally accepted) but it certainly would do so if it were confirmed.
- Modern extensions of this idea propose that all interactions between elementary particles can be accounted for by invoking further wrapped- up dimensions. For example, in 'superstring' theory the total number of dimensions is 10, so one has a 4-dimensional 'external' space and a 6- dimensional 'internal' space. There were originally five different versions of superstring theory, but recent developments

suggest that all of these are part of a more embracing 11-dimensional picture called *M-theory* (where M stands for 'mother' or 'magic' or 'mystery'). A recent variant of this idea suggests that some of the extra dimensions may not be compactified after all but extended. In this case, the physical world can be regarded as a 4-dimensional 'brane' in a higher-dimensional 'bulk'. (This proposal will be discussed in more detail later.)

- The final — and as yet incomplete — paradigm shift is associated with *quantum gravity,* the attempt to unify general relativity and quantum theory. According to this paradigm, the notion of space breaks down on scales less than 10^{-33} cm. It must be regarded, not as a smooth continuum, but as a sort of topological foam. Quantum gravity effects must also dominate whenever classical physics predicts 'singularities' (i.e. points of infinite density), such as arise inside a black hole or at the beginning of the Universe.

This brief history of paradigm shifts shows that the 'ultimate reality' revealed by modern physics is very different from the sort of reality experienced by our normal senses, which only provide a very incomplete picture of the world. Indeed, as emphasized by Arthur Ellison (2002), one can regard successive paradigms as providing a sequence of mental models, each of which is progressively removed from common-sense 'materialistic' reality. Thus atomic theory removes our everyday notion of solidity, relativity theory destroys our intuitive ideas of space and time, quantum theory shows that reality is fuzzy, unification theories reveal dimensions of which we have no direct experience, and quantum gravity goes beyond space and time altogether. Since the ultimate nature of reality can only be appreciated intellectually, it is ironic that many physicists play down the significance of mind. There is also the puzzling feature that the world is understandable by humans at all.

A Brief History of Paraphysics

Physical theories of psi inevitably reflect the physics of their time. An excellent review by Beichler divides the history of the subject into what he terms the pre-scientific, early-scientific, middle-scientific and late-scientific periods. In the pre-scientific era (before 1850) he highlights higher-dimensional theories of spirits (More) and the idea of animal magnetism (Mesmer). These proposals, although primitive, might be regarded as precursors of modern hyperspatial and electromagnetic theories. The early-scientific period (1850-1930) covers the first decades of the SPR and saw attempts to relate psi (including the possibility of survival) to a new force (Crookes), thermodynamics (Tait & Stewart), a fourth dimension (Zollner) and some form of semi-physical 'metetherial' world (Myers). The middle- scientific period (1930-1970) saw developments of these approaches but mainly by non-physicists.

Beichler regards the late-scientific period as starting with the acceptance of the Parapsychological Association into the AAAS in 1970. Four years later, James Beal and Brendan O'Regan declared the emergence of the new science of paraphysics in an influential volume edited by Edgar Mitchell (1974), although the term itself goes back well before that.

This was also the year in which *Nature* published a landmark paper by Russell Targ and Hal Puthoff (1974), describing their investigations into the physical aspects of psi under the auspices of the US-Government- funded 'Stargate' programme.

The same period saw several other edited volumes on the subject as well as the founding of two dedicated journals: *The Journal of Paraphysics* and *Psychoenergetics: The Journal of Psychophysical Systems*. Although these continued to be published into the next decade, progress slowed in the 1980s, when the hopes for a quick-fix theory seemed to fade. The terms 'paraphysics' and 'paranormal' also became tainted by association with the New Age movement, so when Robert Jahn established the PEAR group at Princeton in 1979, he used the term 'anomalous' rather than 'paranormal'.

Connecting ESP and Physics

In deciding whether psi can connect with physics, we first need to decide which Class-3 phenomena we are trying to explain, since some clearly present a greater challenge to theorists than others. This raises the question of whether there are different *levels* of psi, requiring increasing modifications to physics. For example, one might assume that macro-PK poses more of a challenge than micro-PK because more energy is involved; and that precognition is more problematic than clairvoyance, and retro-PK more problematic than PK, because they involve time as well as space displacement. There is also the question of whether mind- mind interactions (such as telepathy) are fundamentally distinct from mind-matter interactions (such as PK) or whether they are aspects of a single unitary phenomenon.

Most paraphysicists would probably agree that one should try to obtain as unified a description of psychic phenomena as possible, without invoking a new feature of physics for each one. This would correspond to the 2/3 boundary progressing in a single big step rather than a lot of little steps. Indeed, the introduction of the single term 'psi' (although very loosely defined) might be thought to anticipate that. In particular, it is important to have a unified description of psi as it appears in the laboratory and in the field. For example, there has been a large amount of laboratory work on micro-PK (the influence of psi on a system which is intrinsically probabilistic), with associated theoretical attempts to explain this in terms of quantum effects. However, there has been relatively little attempt to apply these models to the much more dramatic macro-PK manifestations which arise in (say) poltergeist cases. Indeed, some theorists seem to accept micro-PK but remain sceptical of macro-PK, although one might hope that these phenomena are two extreme forms of a single psychokinetic interaction. A similar dichotomy arises when we consider ESP. In laboratory experiments, we do not usually know which 'hits' are due to chance and which are due to psi—indeed some theorists have argued that no transmission of information need be involved at all (Lucadou, 1995). However, it is hard to see how such a model can be extended to some real-life situations (e.g. crisis apparitions), in which genuine information seems to be conveyed. Likewise, one might hope that presentiment effects observed in the laboratory.

It is important to distinguish physical *theories* of psi from physical *dependencies* of psi or physical *consequences* of psi. As regards the dependencies, various physical influences have been claimed to modify the efficacy of psi—for example, geomagnetic effects or local sidereal time However, this may just reflect the sensitivity of the psychic organ (e.g. some part of the brain) to such influences and may have nothing to do with the mechanism of psi. On the other hand, geomagnetic effects could still be relevant to the mechanism if one attributed psi to extremely low frequency radio waves, a topic well reviewed by Harvey Irwin and Caroline Watt. In the context of micro-PK experiments, there is also the possibility (Stevens) that geomagnetic effects may directly influence the Random Event Generator (REG). Sometimes—as with the claim that sound of a particular frequency can produce apparitions)—it is not clear whether the physical effect is triggering psi or some non-psychic process (like a hallucination). As regards the consequences of psi, ESP may trigger various physical reactions in a subject—such as an electrodermal response—even if this is not recognized consciously. But again this may have nothing to do with the mechanism.

On the other hand, some physical features of psi clearly have important implications for its nature. For example, it has been argued that the presentiment effect may reflect some form of time-symmetry effect (Bierman & Radin, 1997) and a similar idea arises in attempts to link precognition (precall) with memory (recall) (O'Donnell, 2006). This touches on a profound puzzle: even though our conscious experience of the world entails a time-asymmetry, all the equations of physics are time-symmetric. In particular, the solutions of wave equations may involve both 'retarded' and 'advanced' parts (corresponding to propagation along the future and past light-cones, respectively). Although the latter are usually rejected as being acausal (in the sense that they would allow the present to affect the past and the future to affect the present), nothing in known physics precludes them. Indeed, one formulation of electrodynamics explicitly invokes the existence of advanced waves (Wheeler & Feynman,). What is particularly exciting is that quantum experiments now provide a possible way of searching for them (Cramer) and a whole session was devoted to this topic at a recent AAAS meeting. If retrocausal effects were demonstrated, this could have profound implications for psi.

Also of theoretical relevance is the suggestion that the outputs of REGs in micro-PK experiments may contain 'signatures' specific to the individuals trying to influence them (Radin). Indeed, Paul Stevens has designed a special 'signature detection unit' and claims to have found such effects already. Although this interpretation is not completely secure, if it were confirmed, it would support the notion that psi involves the transmission of a *signal.* This could suggest a simpler type of ESP experiment, in which the subject picks up the psychic 'call signal' rather than the message itself

Another interesting issue arises in the context of the sort of micro-PK experiments carried out by the PEAR group). It is usually assumed that micro-PK operates by shifting the mean of a supposedly random distribution, and this seems to be indicated by at least some meta-analyses. On the other hand, the meta-analysis of Fiona Steinkamp gives a much weaker effect.

It should be stressed that the balance effect does not constitute a full theory of psi, since it does not explain how the interaction with consciousness actually arises. The same criticism could be levelled at many other purported theories of psi. This emphasizes that there are different *levels* of explanation. In particular, one must distinguish between physical and psychological levels of explanation. There is a large literature on psychological theories of psi and this is very relevant to its experiential aspects. For example, there has been much interest in whether sensory models (Irwin) or memory models (Roll) best explain ESP, and in the 'psi-mediated instrumental response' model or 'first sight' model. Parapsychologists are also interested in identifying the personality characteristics of subjects who score highly in laboratory experiments. However, none of this may have any bearing on the more fundamental question of how ESP works. Sometimes, of course, it is not clear whether a feature is physical or psychological.

Finally, in producing a physical theory of psi, we need to decide whether we are demanding a new paradigm of physics or merely tinkering with the current one. It is natural to start off by trying the second (less radical) approach, and there are many reviews of 'tinkering' models. However, the danger is that one will end up grafting so many extra bits onto the old paradigm (like adding epicycles to the Ptolemaic model of the Solar System) that it becomes hopelessly complicated. There is also the problem of *testability:* i.there are actually many models for psi and, by adding enough bits to) the standard paradigm, one can doubtless explain anything. Generally speaking, the experimental evidence indicates that ESP can occur at great distances and does not decline with distance. These findings do not fit well with most hypotheses that physical energies mediate the transmission of extrasensory information. Indeed, the information transmission model may itself be erroneous.

However, as discussed below, even if signalling models cannot work in four dimensions, they may still be viable in higher dimensions, since the viewer and the viewed may become contiguous in the higher-dimensional space. This is a crucial feature of my own proposal.

There are also many theories which invoke some form of *biophysical* field, even though the status of such fields is questionable from a physicist's perspective. Mesmer's early ideas on animal magnetism and vitalistic fluids might be included in this category. More recent proposals include biofields, 4-fields (Wasserman), biotonic fields (Elsasser), eidopoic fields (Marshall), psi fields (Roll), bioplasmic fields (Inyushin) and biogravity fields (Dubrov),. Unfortunately, none of these approaches has gained general acceptance among paraphysicists and all of them have been criticized on the grounds that they are *ad hoc* and unfalsifiable. On the other hand, the link with biology is important and reflects the growing interaction between physicists and biologists in orthodox science. It also raises the issue of whether psi is involved in some forms of complementary medicine.

Quantum Models

We have seen that quantum theory — which for present purposes we regard as part of the current paradigm — provides at least some scope for an interaction of consciousness with the physical world. It also completely demolishes our normal concepts of physical reality, so it is not surprising that some paraphysicists have seen in its weirdness some hope for explaining psi. Indeed, E. H. Walker has argued that *only* quantum theory can explain psi:-

This must lie at the heart of the solution to the problem of psi phenomena; and indeed an understanding of psi phenomena and of consciousness must provide the basis of an improved understanding of quantum mechanics.

Jahn also takes this view, arguing that consciousness has two complementary aspects: one particle-like (localized) and the other wavelike (non-localized). However, merely invoking qualitative similarities with quantum effects does not constitute a proper explanation.

The most concrete realization of the quantum approach is 'observational theory', according to which consciousness not only collapses the wave-function but also introduces a bias in how it collapses. In this picture all psi is interpreted as a form of PK which results from the process of observation itself (i.e. there must be some kind of feedback). For example, clairvoyance is supposed to occur because the mind collapses the wave-function of the target to the state reported. This process can even explain retro-PK, since it is assumed that a quantum system is not in a well-defined state until it has been observed. Another feature of observational theory is that the brain is regarded as being akin to an REG. Thus an ordinary act of will occurs because the mind influences its own brain, and telepathy occurs because the mind of the agent influences the brain of the percipient. Of course, there is still the question of *how* consciousness collapses the wave- function. One possibility is to modify the Schrodinger equation in some way (Lucadou & Kornwachs, 1976).

Observational theory has the virtue that it can make *quantitative* predictions. For example, one can estimate the magnitude of PK effects on the basis that the brain has a certain information output (Mattuck) and the results seem comparable with what is observed in macro-PK effects. On the other hand, observational theory also faces serious criticisms. One can object on the grounds that psi sometimes occurs without any feedback. For example, Beloff has pointed out that there are pure clairvoyance experiments in which only a computer ever knows the target. One can also question the logical coherence of explaining psi merely on the grounds that one observes it and there are alternative models for retro-PK. Finally, David Bohm (1986) has cautioned that the conditions in which quantum mechanics apply (low temperatures or microscopic scales) are very different from those relevant to the brain.

Nevertheless, many paraphysicists back some form of quantum approach Some proposals exploit the non-locality of quantum theory, as illustrated by the famous EPR paradox (Einstein, Podolosky & Rosen,). An atom decays into two particles, which go in opposite directions and must have opposite (but undetermined) spins. If at some later time we measure the spin of one of the particles, the other particle is forced instantaneously into the opposite spin-state, even though this violates causality. This non-locality effect is described as 'entanglement' and Bohm tried to explain this in terms of hidden variables, which he invoked as a way of rendering quantum theory deterministic. Experiments later confirmed the non-locality prediction (Aspect) and thereby excluded at least some models with hidden variables (though not Bohm's). Indeed, John Bell, who played a key role in developing these arguments (Bell) and was much influenced by Bohm's ideas, compared the non-locality property to telepathy. Einstein made the same comparison, although he intended it to be disparaging!

Although quantum entanglement has now been experimentally verified up to the scale of macroscopic molecules, it must be stressed that it is not supposed to allow the transmission of *information* (i.e. no signal is involved). For example, attributing remote viewing to this effect would violate orthodox quantum theory. Theorists have reacted to this in two ways. Some have tried to identify what changes are necessary in quantum theory in order to allow non-local signalling (Valentini).. More generally, Jack Sarfatti (1998) has argued that signal non-locality could still be allowed in some form of 'post-quantum' theory which incorporates consciousness. He regards signal locality as the micro-quantum limit of a more general non-equilibrium macro-quantum theory (cf. Bohm & Hiley, 1995). The relationship between micro and macro quantum theory is then similar to that between special and general relativity, with consciousness being intrinsically non-local and analogous to curvature. His model involves non-linear corrections to the Schrodinger equation and may permit retrocausal and remote viewing effects (Sarfatti,).

Others accept that there is no signalling but invoke a 'generalized' quantum theory (Atmanspacher, Romer & Walach) which exploits entanglement to explain psi acausally. This is also a feature of the model of pragmatic information (Lucadou & Kornwachs), which interprets psi effects as meaningful non-local correlations between a person and a target system. This model may account for many of the observed features of psi, including the difficulty of replicating psi under laboratory conditions (Lucadou, Romer & Walach. It may also be relevant to homeopathy (Walach).

Radin has argued that entanglement is fundamental to psi. This is because he regards elementary-particle entanglement, bio- entanglement (neurons), sentient-entanglement (consciousness), psycho- entanglement (psi) and socio-entanglement (global mind) as forming a continuum, even though there is an explanatory gap (and sceptics might argue an evidential gap) after the second step. If the Universe were fully entangled like this, he argues that we might occasionally feel connected to others at a distance and know things without use of the ordinary senses. This idea goes back to Bohm (1980), who argued that there is a holistic element in the Universe, with everything being interconnected in an implicate order which underlies the explicit structure of the world:- The essential features of the implicate order are that the whole Universe is in some way enfolded in everything and that each thing is enfolded in the whole. This implicit order is perhaps mediated by psi (Pratt, 1997). Most mainstream physicists regard such ideas as an unwarranted extension of standard quantum theory, but one clearly needs some sort of extension if one wants to incorporate mind into physics.

There are various other quantum-related approaches to explaining psi. Some of these exploit the effects of 'zero point fluctuations' (Puthoff, 1989) or 'vacuum energy' (Laszlo, 1993). This is a perfectly respectable physical notion, so it is not surprising that some people have tried to relate this to the traditional metaphysical idea that there is some all- pervasive energy field which connects living beings (eg. chi, qi, prana, elan vital). Indeed, Puthoff (2007) views the zero-point-energy sea as a blank matrix upon which coherent patterns can be written. These correspond to particles and fields at one extreme and living structures at the other, so some connection with psi is not excluded. He writes:-

All of us are immersed, both as living and physical beings, in an overall interpenetrating and interdependent field in ecological balance with the cosmos as a whole, and even the boundary lines between physical and metaphysical would dissolve into a unitary viewpoint of the Universe as a fluid, changing, energetic information cosmological unity. A related proposal is that the radiation associated with zero-point-energy might be identified with 'subtle energy fields' (Srinivasan, 1988). These allegedly involve some form of unified energy of such low intensity that it cannot be measured directly (Tiller, 1993). In the electromagnetic context, this idea was introduced to describe the quantum potential (Aharonov & Bohm, 1959) and maybe relevant to Bohm's implicate order.

Although these ideas might be regarded as being on the fringe of the standard paradigm, the recent discovery that 70% of the mass of the Universe is in the form of 'dark energy'— most naturally identified with vacuum energy — is stimulating interest in this sort of approach. For example, Sarfatti has a model which associates both consciousness and dark energy with the effects of vacuum fluctuations, although he does not explicitly identify them.

It should be cautioned that the literature in this area comes from both expert physicists and non-specialist popularizers, so it is important to discriminate between them (Clarke & King, 2006). Although quantum theory is likely to play some role in a physical model for psi, my own view is that a full explanation of psi will require a paradigm which goes beyond standard quantum theory. Of course, nobody understands quantum theory anyway, so claiming that it explains psi is not particularly elucidating—it just replaces one mystery with another one (Clarke, 1996). Also, many of the above proposals already deviate from standard quantum theory, so this raises the question of how radical a deviation is required in order to qualify as a new paradigm. In my view, most of those mentioned above are insufficiently radical and one needs a new approach —perhaps of the kind envisaged by Bohm—that can explain *both* psi *and* quantum theory. One also suspects that the new paradigm will incorporate the idea of retrocausality discussed earlier, since proposed tests of this all involve some form of EPR effect (Cramer).

Higher –Dimensional Models

The space perceived by our ordinary senses is clearly 3-dimensional. However, even before relativity theory introduced the idea of time as a 4th dimension, it was popular to invoke an extra dimension of space as an explanation of paranormal phenomena (Rucker). Indeed, Henry More's book *Enchiridion Metaphysicum* associated spirits with a 4th dimension as early as 1671, although his contemporary John Wallis regarded this as "a monstrous invention, less likely than a chimera". Less mystical treatments of the 4th dimension—some of which were later influential in the development of general relativity—were presented in the 19th century by such eminent mathematicians as Moebius, Gauss, Riemann, Helmholtz and Clifford. However, the mystical connection resurfaced in what Beichler terms the early-scientific period, when the astronomer Johann Zollner (1880) invoked a 4th dimension in order to explain some of the spiritualistic phenomena of the medium Henry Slade. Unfortunately, his career was destroyed when Slade was later revealed to be a fraud.

The mystical implications of an extra dimension were explored further (at least by analogy) in 1884 when Edwin Abbott described the effects of a 3rd dimension on the inhabitants of a 2-dimensional world in his book *Flatland* (1983). The idea of a 4th dimension was also championed by Charles Hinton, who coincidentally worked in a patent office in Washington at the same time as Einstein worked in one in Berne. His 1880 book *What is the 4th Dimension ?* was subtitled *Explaining Ghosts,* and his 1885 book *Many Dimensions* explicitly claimed that minds extend in the 4th dimension (Hinton, 1980). Although Abbott and Hinton were really popularizers of ideas developed earlier, they were very influential in generating public interest in the topic. Indeed, the period from 1890 to 1905 was a golden age for the 4th dimension and spirits in unseen hyper- space were particularly popular with clergymen. A. T. Schofield's *Another World* (1888) put God in the 4th dimension, while Arthur Willink's *The World of the Unseen* (1892) put Him in an infinite-dimensional space! The idea also influenced literature: Oscar Wilde's *The Canterville Ghost* (1891) lampooned the 4th dimension, while H. G. Wells's *Time Machine* (1895) presaged the idea of time as a 4th dimension.

Most of these ideas fell by the wayside after Einstein formulated his theory of special relativity in 1905. Einstein showed that the 4th dimension really does exist but that it is time and not appropriate for the more exotic purposes envisaged above. However, in the following decades, there were various attempts — mainly from non-physicists — to use Minkowski space (or some extension of it) to explain psi. For example, this featured prominently in the theosophical tradition: P. D. Ouspensky associated the 4th dimension with mystical unity as early as 1908 and developed this idea further in subsequent works. A book by Whateley Smith (1920)—later Whateley Carington —associated survival with Einstein's 4th dimension, while J. W. Dunne) introduced extra time dimensions (an infinite number of them) to explain the flow of consciousness and dream precognition. His approach was later discredited by C. D. Broad (1953), who nevertheless introduced his own model with just two times. There were also attempts to link the mind with extra dimensions

In the late scientific period, more mathematical models have been offered by physicists themselves. In particular, a series of papers have studied 8-dimensional models, in which one complexifies the four coordinates of space and time. Indeed, this model has also been proposed in standard relativity, as a way of unifying the equations of Newton, Maxwell, Einstein and Schrodinger (Newman, 1973). The paraphysical application of this idea seems to have been proposed independently by Russell Targ et al. (1979) and Elizabeth Rauscher (1979, 1983), and has recently been reviewed by them (Rauscher & Targ, 2001, 2002). On the other hand, Michael Whiteman (1977) has invoked a 6-dimensional model, with three real times, claiming that this incorporates the Maxwell and Dirac equations. A similar model has been proposed by Burk- hard Heim and this is alleged to explain elementary particle masses (Auerbach & Ludwiger,) although variants of this model allow up to 12 dimensions. There is also a 12-dimensional model with three complex space and three complex time dimensions (Ramon & Rauscher, 1980). All these extensions of relativity theory suppose that points can be contiguous in some higher-dimensional space even though they are separated in ordinary spacetime. This contiguity is supposed to explain how events at remote locations or times can be present in consciousness. Indeed, this idea goes back to Gertrude Schmeidler, who invoked folding in higher dimensions to explain some features of psi.

A rather different higher-dimensional approach is to invoke an extra *spatial* dimension. The idea of a 5th dimension was introduced within physics in the 1920s by Theodor Kaluza and Oskar Klein in their attempt to provide a geometrical description of the unification of gravity and electromagnetism. It did not attract much attention at the time but, in principle, a 5th dimension can take on the same role as that attributed to the 4th dimension in the pre-relativistic period. For example, there have been attempts to use a 5th dimension to link parapsychology with UFOs (Brunstein) and with the manipulation of spacetime curvature by 'biogravity' fields. John Ralphs claims that (what he terms) a 4th dimension (but is really a 5th dimension) can explain such diverse phenomena as spirit communications, movements of objects through space and time, clairvoyance and dowsing. More recently, Julie Rousseau has revived Zollner's proposal that phenomena such as teleportation, apportation and materialization could result from interactions in the 5th dimension. The strongest advocate of the 5-dimensional model of psi is Beichler and his approach is discussed in more detail later. He has also provided an excellent account of the history of the 5th dimension in orthodox physics (Beichler).

By invoking enough dimensions, one can doubtless explain anything. However, like the biofield models, these proposals are subject to the criticism that they are unfalsifiable and do not make quantitative predictions. Also, like the quantum models, they come from a combination of specialist physicists and non-specialist popularizers, so discrimination is required in assessing them. Nevertheless, as a result of the rising interest in Kaluza-Klein theory, there is no denying that the idea of extra dimensions has now taken centre-stage in modern physics. Although there is a debate within the physics community as to whether these higher dimensions have physical significance or are just a mathematical artefact (Woit), they are of obvious interest to paraphysicists.

THE NEED FOR A NEW PARDIGM

Although all of the above approaches may be relevant to a final theory of psi, my own view is that a full explanation will require a paradigm shift which goes beyond them and perhaps combines them in some way. Even though some psychic phenomena might be amenable to explanation within the current paradigm, I do not believe all of them can be, so the new paradigm must incorporate psi in a more embracing way than any of the particular models described above. Beichler also takes this view:-

Parapsychology is littered with hypothetical structures to explain the psi process, or some features of it, but no comprehensive theory... . Either psi does not exist or psi is so fundamental that it is intimately interwoven into the very fabric of reality.

Indeed, he argues that the conditions for a paradigm shift are already present and that psi could be the catalyst in promoting it. But what sort of paradigm shift would be required to accommodate psi? In Beichler's view, it must go beyond quantum theory and general relativity, since both of these are semi-classical, in the sense that they inherit some features of the classical paradigm. (He argues, rather perversely, that quantum theory inherits more classical features than general relativity.) I would agree with this, although it is hard to believe that quantum theory—or at least some deeper theory on which it is based—will not play some role. Whatever the new paradigm, it is likely to be sufficiently radical to provoke a dispute over whether it should be classed as physics, which is why the term 'hyperphysics' may be more acceptable.

One ingredient of the new paradigm may be a transcendence of the usual ideas of space and time. Indeed, as illustrated by the previous quotation from Palmer, many people reject physicalistic models at the outset precisely because psi seems to exhibit this property. However, this rejection is premature since the transcendence of space and time already arises in physics itself in the context of quantum gravity. Indeed, it is possible that the long-sought unification of relativity and quantum theory will itself play a role in the new paradigm. The proposal that quantum gravity may be relevant to the collapse of the wave-function already hints at this (Penrose). The transcendence of space and time also arises in the contexts of more radical proposals: for example, in Rupert Sheldrake's model of 'formative causation' and Jung and Pauli's model of 'acausal synchronicity' Another crucial ingredient of the new paradigm must presumably be consciousness, although it should be stressed that many psi processes maybe *unconscious* and Beichler has argued that'life' is the more relevant ingredient. We have seen that there is some indication from physics itself that consciousness is a fundamental rather than incidental feature of the Universe. This idea arises explicitly in the writings of Bohm (1980), who argues that quantum theory introduces a mind-like quality into the Universe. In his holographic model, there is a unity of consciousness, a greater collective mind with no boundaries of space or time:—

All this implies a thorough-going wholeness, in which mental and physical sides participate very closely in each other. Thus, there is no real division between mind and matter, psyche and soma.

Bohm achieves this by introducing a quantum 'superpotential' (corresponding to a new sort of force) and this also allows an organizing principle, which is rather similar to Sheldrake's morphogenetic field. But how can one incorporate mind into the picture explicitly? In the next part, I will argue that the invocation of higher dimensions (i.e. more than the four usually envisaged) is necessary. However, it is not the *existence* of the higher dimensions which constitutes the new paradigm—that is already accepted by string theorists—but rather the association of those higher dimensions with mind.

This is not a new idea, since we have seen that people have long attributed mental experiences to higher dimensions. For example, Whiteman has argued that the whole domain of mystical experience can be accommodated in a higher-dimensional approach, and Heim (1988) has made similar claims.. A number of physicists have emphasized this, but perhaps the most mathematically sophisticated attempt to connect matter and consciousness through these developments comes from Saul-Paul Sirag. The key to his approach is group theory: he associates the hierarchy of consciousness with the hierarchy of what mathematicians term 'reflection spaces'. In particular, an important role is attributed to 7-dimensional reflection space, which is a symmetry group of one of the Platonic solids.

Chapter 23

BRAIN WAVES AS THE SOLUTION OF THE MODIFIED SCHRODINGER EQUATION

QUANTUM MECHANICS AND CONSCIOUSNESS

A great deal has been published in recent years about the brain and consciousness. The philosopher David Chalmers, who specializes in questions of consciousness, has written an excellent overview of all the different theories about the brain-consciousness relationship. He starts by describing three materialist and reductionist models, A, B, and C. The first model (A), which he labels "monistic materialism," is based on the premise that everything is matter. Because the brain is made up of neurons undergoing physical and chemical processes, adherents of this theory believe that by explaining these processes in the brain they can also explain consciousness. The most commonly heard interpretation of this model posits that consciousness is merely an illusion. The second materialist model (B) is based on the premise that consciousness must be identical to processes in the brain because in a functioning brain there is a link between certain activities in the brain and certain experiences of consciousness. Adherents of the third model (C) admit tat consciousness cannot be reduced to brain function just yet but believe that with scientific progress this will only be a matter of time.

Chalmers presents a detailed case against these three materialist approaches. His first counterargument is that while the structures and functions of the brain can be explained, this in itself is not enough to explain consciousness. His second counterargument involves zombies, imaginary creatures that are physically identical to human beings but lack a human consciousness. If zombies are a theoretical possibility, their brain function must be identical to the human brain, in which case the absence of consciousness in these fictional creatures means that consciousness is immaterial. His third antimaterialist argument invokes known facts about consciousness that cannot be explained on the basis of physiological brain activities. In theory, scientists could know everything there is to know about brain function and still not be familiar with all aspects of consciousness. Even with complete materialist knowledge, Chalmers suggests, we cannot know everything there is to know about consciousness.

Chalmers follows this with a description of three nonreduction-ist and immaterial models, D, E, and F, furnishing each model with commentaries from proponents and opponents. The fourth model (D) describes the "interactionist-dualism" developed by the Nobel Prize winner and neurophysiologist John Eccles and philosopher of science Karl Popper on the basis of the radical dualism of the sixteenth-century mathematician and philosopher Rene Descartes. Consciousness and the brain are radically different yet somehow highly interactive. According to Chalmers, this model is seen as incompatible with *classical* physics, whereas concepts from *quantum* physics, such as the collapse of probability waves caused by a deliberate observation or measurement, could actually support this model. As outlined in the previous chapter, not all quantum physicists accept the role of consciousness in quantum physics. Chalmers notes that philosophers usually reject interactionism with arguments from quantum physics, while physicists tend to reject the model on philosophical grounds (dualism). The fifth explanatory model (E) for the mind-brain relationship is called "epiphenomenalism" or "weak dualism," which posits that certain areas of brain function trigger certain experiences of consciousness, but that consciousness has no effect on brain or bodily function. This concept resembles the materialist vision. Consciousness is said to be the effect of chemical and electrical processes but cannot actually influence these processes. If so, the experience of pain could never cause a physical reaction, and people could never decide to take action. Neuroplasticity also argues against this model given that empirical studies have shown that the mind is capable of permanently changing the anatomy and function of the brain, as discussed earlier.

As his sixth and final model (F), Chalmers cites "phenomenalism" or "immaterial (or neutral) monism." This model is also known as "panpsychism" or "idealism." According to this model, all material, physical systems contain a form of subjective consciousness at an elementary or fundamental level, and all matter has phenomenal properties (that is, properties based on subjective observation). This model posits that consciousness has a primary presence in the universe and that all matter possesses subjective properties or consciousness. In this model consciousness is not only an intrinsic property of all matter, but physical reality is even formed by consciousness. Chalmers thus ascribes consciousness a distinctly causal role in the physical world. He mentions another theoretical possibility in which the intrinsic properties of the physical world *are* not phenomenal properties but rather *possess* phenomenal properties, which makes them protophenomenal properties. For this reason he prefers the term *panprotopsychism.*

Not everybody will agree with the latter model, in which all matter possesses subjective properties or consciousness; most people will favor a materialist model to explain the brain-consciousness relationship. But Chalmers appears to support panprotopsychism and believes that it merits further study.

Empirical research into human consciousness during a loss of all brain function (NDE) also seems to corroborate model F. The content of an NDE suggests that consciousness may be nonlocal. When brain function is impaired, NDErs experience an enhanced consciousness detached from the body followed by a conscious reentry into the body, rendering a materialist explanation of consciousness highly unlikely. When the brain functions normally, an NDE with an out-of-body experience can be triggered by mortal fear or stress while spontaneous out-of-body experiences are not uncommon at a young age. Neuro-physiological studies have shown that brain activity cannot account for the content of thoughts and feelings whereas there is incontrovertible evidence for the mind's influence on the brain, given that the anatomical structure of the brain and its associated functions can change in response to experiences in the mind (neuroplasticity).

The Materialist Approach

A majority of contemporary Western scientists specializing in consciousness research, such as neuroscientists, psychologists, psychiatrists, and philosophers, espouse a materialist and reductionist explanation for consciousness (model A, B, or C). The well-known philosopher Daniel Dennett, for example, adheres to model A, monistic materialism.[3] As I wrote in the introduction to this book, Dennett, like many others, is of the opinion that consciousness is nothing but matter and that our subjective experience of consciousness as something purely personal and distinct from other people's consciousness is merely an illusion. According to Dennett, consciousness is produced by the matter that comprises our brain. This materialist hypothesis is supported by scientific patterns of thought and paradigms that he and many other scientists and philosophers deem absolutely unassailable and are therefore reluctant to challenge. Scientists often struggle to free themselves from prevailing paradigms. And such dogmatic convictions seem to beget prejudice. It prompted Albert Einstein to say, "It is harder to crack a prejudice than an atom."

If the materialist standpoint were correct, everything we experience in our consciousness would be nothing but the expression of a machine controlled by classical physics and chemistry. In Dennett's view, our behavior is the inevitable result of neuronal activity in our brains. The idea that all thoughts and feelings are no more than a consequence of brain activity obviously means that free will is an illusion. In response to this materialist position I cite neurophysiologist John C. Eccles:

> I maintain that the human mystery is incredibly demeaned by scientific reductionism, with its claim in promissory materialism to account eventually for all of the spiritual world in terms of patterns of neuronal activity. This belief must be classed as a superstition... . We have to recognize that we are spiritual beings with souls existing in a spiritual world as well as material beings with bodies and brains existing in a material world.[4]

The materialist approach, which is based on the premise that consciousness is a product or effect of brain function, is taught at many medical schools in the Western world. The approach is generally not made explicit and simply taken for granted without any kind of debate. Not surprisingly then, nearly all Western doctors believe that consciousness is the result of brain function. I myself am the product of an academic environment and was taught that there is a reductionist and materialist explanation for everything. I always blindly accepted this perspective, not just as a medical student and doctor, but also as the son of a neurologist. According to the materialist approach, the experience of consciousness during a spell of unconsciousness, cardiac arrest, coma, or a period of brain death is of course impossible. If, citing the results of empirical studies of NDE, somebody hints at the possibility of consciousness at a moment when all brain function has ceased, this is usually rejected as unscientific. Such a response from the scientific community is not new. Here is a quote from Dutch author and psychiatrist Frederik van Eeden from 1894:

> The main concern is that the claims of a few scientists are fiercely disputed by most of their peers, not on the basis of research, but a priori; not even with rational arguments, but with emotional motives. Emotional motives with their aftermath of ridicule, contempt, and

insinuation, based solely on an unphilosophical attachment to a closed system. It seems barely credible

Sadly, the scientific community has changed little over the past hundred years.

NEAR-DEATH EXPERIENCE, CONSCIOUSNESS AND THE BRAIN

What have we read about the relationship between consciousness and the brain in the previous chapters?

- Many serious and trustworthy people have reported that, to their great surprise, they were able to experience an enhanced consciousness, independently of their body.
- On the basis of a few scientifically sound studies of NDE among cardiac arrest survivors, researchers have come to the conclusion that current scientific knowledge cannot offer an adequate explanation for the cause and content of a near-death experience.
- Some prospective, empirical studies provide conclusive evidence that it is possible to experience an enhanced and lucid consciousness during a cardiac arrest.
- We appear to have scientific proof that the cerebral cortex and brain stem are devoid of measurable activity during a cardiac arrest arid that the clinical picture also reflects a loss of all brain function.
- Brain studies have shown that under" normal circumstances a functioning, collaborative network of brain centers is a prerequisite for the experience of waking consciousness. This is absent during a cardiac arrest.
- Oxygen deficiency in itself provides no explanation because NDEs can be reported under circumstances that are not life-threatening, such as mortal fear or a serious depression.
- Our mind is capable of altering the anatomy and function of the brain (neuroplasticity).
- In many respects, both consciousness and brain function remain a huge mystery.

Some prospective and many retrospective studies of near-death experience have shown that various aspects of an NDE correspond with or are analogous to some of the basic principles from quantum mechanics, such as nonlocality, entanglement or interconnectedness, and instantaneous information exchange in a timeless and placeless dimension. Past, present, and future are everywhere at once (nonlo-cally). Earlier I outlined a few generally accepted principles of quantum physics because I am convinced that these are essential to our understanding of the brain-consciousness relationship. In my view, the quantum physics idea that consciousness determines if and how we experience our reality is particularly important for the further theoretical underpinning of this relationship. However, this radical interpretation of quantum physics is not yet commonly accepted.

The Continuity of Consciousness

Science challenges us to devise, test, and discuss new ideas that might explain the reported connection between one's own consciousness and that of other living persons or deceased relatives. The same applies to nonlocal phenomena such as the life review and preview, in which past, present, and future can be experienced simultaneously and which elude our conventional embodied conception of time and space. For me the biggest challenge is to find an explanation for the fact that an enhanced consciousness can be experienced independently of the body during the temporary loss of all cortical and brain-stem function.

A final theoretical possibility, one that has not been mentioned so far, is the theory of transcendence, or rather the continuity hypothesis. It views the NDE as an altered state of consciousness in which memories, self-identity, lucid thought, and emotions can be experienced independently of the unconscious body and in which (extrasensory) perception outside the body remains a possibility. The Dutch study and other empirical studies have shown that NDErs can experience an enhanced consciousness independently of their normal, embodied waking consciousness. I am reluctant to use the word *transcendence* because it Suggests something transcending or rising above the body. Transcendence is usually associated with the supernatural or with the concept of; transcendental meditation; hence my preference for the term *continuity hypothesis*. Besides, because consciousness is continuous and nonlocal, if! do not believe that consciousness rises above the body. It is always present outside and often inside the body. This chapter will shed further light on the concept of the continuity of consciousness.

New Scientific Concepts

As mentioned, current medical and scientific knowledge cannot account for all aspects of the subjective experiences reported by cardiac arrest survivors with an NDE. However, I believe that science means asking questions with an open mind. And science is also about searching for possible explanations for new, initially perplexing problems instead of clinging to old facts and concepts. The problem lies less in accepting the content of new ideas than in rejecting old and familiar conceptions. The history of science tells us that sooner or later—and sometimes very soon—new empirical findings will force us to abandon our acquired knowledge. Quantum physicist David Bohm believed that "fixed ideas which underlie scientific hypotheses are not aids but obstructions to clarity, and that a methodology which combines discipline with openness would be better equipped to keep pace with the truth that is revealed as scientific investigation progresses and deepens. American philosopher of science Thomas Kuhn, claimed that contrary to popular belief, most typical scientists are not objective and independent thinkers. This is a generalization, of course, but he believed that scientists tend to be rather "conservative individuals who accept what they have been taught and apply their knowledge to solving the problems that their theories dictate." Most scientists try to reconcile theory and fact within the accepted paradigm, which Kuhn describes as essentially a collection of "articles of faith shared by scientists." All research results that cannot be explained by current scientific theories are labeled "anomalies" because they threaten the existing paradigm and challenge the expectations raised by such paradigms. Needless to say, these findings are initially overlooked, ignored, rejected as aberrations, or even ridiculed. Near-death experiences are such anomalies because their cause and content cannot be accounted for with current medical and scientific ideas about the various aspects of human consciousness and the mind-brain relationship. I believe that anomalies can make a vital contribution to the critical reassessment and, where necessary, rejection of old concepts in favor of new and better theories that do explain these anomalies. In the past anomalies have always been the key to scientific paradigm shifts, just as the initially inexplicable behavior of heated metal prompted the development of quantum physics.*A New Perspective on Consciousness and the Brain* I developed the following views in response to the commonly reported experiences of an enhanced consciousness during a cardiac arrest. This enhanced consciousness features nonlocal aspects of intercon-nectedness, such as memories from earliest childhood up until the crisis that caused the NDE and sometimes even visions of the future. It offers the chance of communication with the thoughts and feelings of people who were involved in past events or with the consciousness of deceased friends and relatives. This experience of consciousness can be coupled with a sense of unconditional love and acceptance while people can also have contact with a form of ultimate and universal knowledge and wisdom.

In this new approach, complete and endless consciousness with retrievable memories has its origins in a nonlocal space in the form of indestructible and not directly observable wave functions. These wave functions, which store all aspects of consciousness in the form of information, are always present in and around the body (nonlocally). The brain and the body merely function as a relay station receiving part of the overall consciousness and part of our memories in our waking consciousness in the form of measurable and constantly changing electromagnetic fields. In this view, these electromagnetic fields of the brain are not the cause but rather the effect or consequence of endless consciousness.

According to this concept, our brain can be compared to a television set that receives information from electromagnetic fields and decodes it into sound and vision. Our brain can also be compared to a television camera, which converts sound and vision into electromagnetic waves, or encodes it. These electromagnetic waves contain the essence of all information for a TV program but are available to our senses only through a television camera and set. In this view, brain function can be seen as a transceiver; the brain does not produce but rather facilitates consciousness. And DMT or dimethyltryptamine, which is produced in the pineal gland, could play an important role in disturbing this process, as we saw earlier. Consciousness contains the seeds of all the information that is stored as wave functions in nonlocal space. It transmits information to the brain and via the brain receives information from the body and the senses. That consciousness affects both form and function of the brain and the body has been described in the discussion of neuroplasticity ("The mind can change the brain"). This view corresponds with what David Bohm has written: "Consciousness informs and in-forms.

Nonlocal Consciousness in Nonlocal Space

My term for the wave functions in nQnlocal space, which possess both personal and universal information, is *nonlocal consciousness.* According to this approach, consciousness has no material basis. I already outlined this vision in 2004 and 2006, but the terminology I now use is slightly different. In this model nonlocal space is more than a mathematical description; it is a metaphysical space in which consciousness can exert influence because nonlocal space possesses subjective properties of consciousness. In this view consciousness is nonlocal and functions as the origin or basis of everything, including the material world.

As I explained in the previous chapter, observation is, by definition, impossible in nonlocal space because everything is based on probability fields (wave functions); time and distance play no role either. In other words, the physical aspect of our consciousness in the material world, which we experience as waking consciousness and which can be compared to the particle aspect of light, stems from the wave aspect of the "complete" and "endless" consciousness created by collapse of the wave function in nonlocal space. This particle aspect, the physical effect of our waking consciousness, is observable and demonstrable in the brain through EEG, MEG, fMRI, and PET-scan technology whereas consciousness in nonlocal space is not directly demonstrable on (quantum) theoretical grounds: everything that is visible emanates from the invisible. To better understand this immeasurable and invisible nonlocal consciousness, one could think of gravity. "While gravity itself is not directly demonstrable or provable, its physical effects certainly are. How can this theoretical approach help us understand the possibility of experiencing consciousness, complete with memories and occasional glimpses of the future, during an NDE when the brain has stopped functioning? It may be helpful to compare this continuous, invisible, and instantaneous interaction between the mind and body to modern worldwide communication. Time and distance appear to play no role in the nonstop global exchange of information because of all the electromagnetic information waves for the ubiquitous cell phone, television, radio, and computer technologies* that surround and penetrate us at all times. These information waves propagate at the speed of light. We are not aware of the hundreds of thousands of telephone calls, hundreds of television and radio broadcasts, and the billions of Internet connections around us day and night, passing through us and through the walls, including those of the room in which you are reading this book. We are not aware of these electromagnetic information waves until we switch on our cell phone, TV, radio, or laptop. But what we receive is not actually in the,appliance. The voice we hear through the receiver is not inside the telephone. The images and music of the TV broadcast are not inside the TV set, and the concert is not inside the radio. We only see and hear the program when we switch on a TV set, and when we switch it off again we stop seeing and hearing it even though the broadcast continues. When we switch on another TV set, we receive the same program again. The connection appears to be nonlocal, and in actual fact all electromagnetic information is disseminated at the speed of light.

We can also compare endless and nonlocal consciousness with the Internet, which does not originate in the computer but is received and made visible to the senses by the computer. Akin to the brain's role in consciousness, a computer has a facilitating function: with the right access codes, a computer allows us to access more than a billion different Web sites. The computer does not produce the Internet any more than the brain produces consciousness. The computer allows us to add information to the Internet just like the brain is capable of adding information from our body and senses to our consciousness. Like a computer, the brain functions as a transceiver. As soon as you switch your computer off, you lose access to all those Web sites. Yet the sites themselves remain available worldwide, in Australia, Africa, Europe, Asia, as well as North and South America. And so it is with consciousness. It is always present. During life, we can experience aspects of consciousness in our body as our waking consciousness. Life allows us to make the transition from nonlocal space to our physical world, space-time. The oxygen deficiency brought on by the stopping of the heart temporarily suspends brain function, causing the electromagnetic fields of our neurons and other cells to disappear and the interface between consciousness and our physical body to be disrupted. This creates the conditions for experiencing the endless and enhanced consciousness outside the body (the wave aspect of consciousness) known as an NDE: the experience of a continuity of consciousness independent of the body. This concept of an enhanced and nonlocal consciousness can account for all elements of an NDE. When the body dies, consciousness can no longer have a particle aspect because all brain function is permanently lost. Endless (nonlocal) consciousness, however, will exist forever as wave functions in nonlocal space.

Scientific Proof of the Nonlocal Entanglement of Consciousness

Experiments appear to provide scientific proof of the nonlocal entanglement or connectedness of consciousness. Pairs of people were placed in two separate Faraday cages, which are rooms shielded from electromagnetic radiation to block out any electromagnetic information transfer. If these two people were strongly connected to each other, such as parent and child or two people who practiced many years of joint meditation, simultaneous changes in their EEG could be registered. In one isolated Faraday chamber, sensory stimulation through random computer-generated flashes of light caused visual evoked potentials in the EEG registration of the stimulated person, and this activity was instantaneously received by the other, unstimulated person in the second Faraday cage. As a result, the registered patterns in the EEG of the unstimulated person changed the moment the lights flashed in the other Faraday cage. This transferred electrical activity, the so-called transferred potentials, the coherence or correlation between the two EEGs, can be ascribed only to nonlocal influence. Because the experiment design excluded electromagnetic information transfer, this correlation cannot be explained with classic scientific models,

Physicist Fred H. Thaheld has outlined a potential scientific basis for this macroscopic and biological nonlocal entanglement. The first studies of this nonlocal entanglement of consciousness were carried out at the University of Mexico by the neurophysiologist Jacobo Grinberg-Zylberbaum. The research initially met with criticism because of its poor design, but scientists at three different laboratories later replicated identical EEG correlations. Two fMRI studies found evidence of nonlocal entanglement between the brains of two isolated individuals while nonlocal influence has also been identified in subjects whose fMRI registration changed significantly when a healer at some distance focused attention on these subjects. And a recent study using laser stimulation and local EEG registration has shown nonlocal biological and macroscopic entanglement between two cultivated specimens of fully isolated human neural networks

All of these carefully executed and replicated empirical studies confirm the nonlocal properties of consciousness and point to a nonlocal entanglement in biological and macroscopic systems such as the brain. Neither the classical physics model of science nor contemporary biological theories can account for this correlation of biological systems. Whether or not quantum theory is capable of doing so is a question I will try to answer in the next few sections.

The Interface between Nonlocal Consciousness and the Brain

The human brain is an extremely complex and in many respects mysterious organ with physiological, chemical, and biological properties. But because consciousness is not physiological, chemical, or biological, the brain is much harder to analyze. Mathematician and physicist Roger Penrose has argued that on theoretical grounds consciousness cannot be produced by the brain. He has also demonstrated that computers will never be able to fully replicate or produce consciousness.

We believe that while quantum physics cannot explain the origins of our consciousness, nonlocal consciousness does have a lot of common ground with widely accepted concepts from quantum physics. In my opinion, quantum physics can help us understand the transition from consciousness in nonlocal space to embodied waking consciousness in our physical, visible world. The aforementioned nonlocal entanglement of consciousness in biological and macroscopic systems, which has been demonstrated by the instantaneous information transfer between tjie brains of two separated subjects resulting in identical EEG and fMRI patterns, can be considered as an initial contribution to explain the transition from aspects of nonlocal consciousness to the brain.

Theories Addressing the Transition from Nonlocal Consciousness to the Physical Brain

The following feature technical descriptions of three different interface or place-of-resonance models that may be able to explain the transition from nonlocal consciousness to the physical brain.

All three are complementary models in which subjective conscious experiences and their corresponding objective, physical brain activities are two fundamentally different manifestations of the same underlying nonlocal reality that cannot be reduced to one another. It is impor-, tant to realize that, in keeping with current interpretations of quantum physics, all three models see the electromagnetic fields of the brain not as the cause but as the effect or consequence of consciousness. The {three theories on the interface or place of resonance are, respectively, the link between nonlocal consciousness and (virtual) photons; the in-jjBuence of nonlocal consciousness on the brain via the quantum Zeno effect; and the nonlocal information transfer from consciousness via iiquantum spin correlation.

How the exact transition ("place of resonance") from nonlocal space to the physical world comes about is not known. In fact, the process will probably never be fully knowable or verifiable. The potential role of DMT in establishing or disrupting this transition or interface will be equally difficult to prove. This means that we will probably never have any experimental evidence for the actual transition or interface between consciousness and the brain. Quantum physics allows for several theoretical possibilities, which are all speculative to a certain degree—fundamentally difficult to prove or disprove. In the previous chapter I discussed a few quantum-mechanical concepts for the transition from consciousness to the brain. Of the following three theories, my personal preference goes to the third, although I believe that all three models are a genuine possibility and in some way complement one another. In the near future, these three models will have to be researched and developed in greater detail.

THE LINK BETWEEN CONSCIOUSNESS AND (VIRTUAL) PHOTONS

Consciousness is nonlocal, that is, everywhere in nonlocal space and intrinsically entangled with all potential information stored in wave functions. Consciousness triggers collapse of the wave function and is thus the source of embodied waking consciousness. There is a theoretical possibility that consciousness in nonlocal space is linked to—or serves as the basis for—the electromagnetic field connected to the nervous system and the brain. In that case consciousness would be hitchhiking, as it were, on the electromagnetic field that probably originates, like consciousness, in nonlocal space. As we saw earlier, neurobiologist Herms Romijn developed this hypothesis, which is based on coherent systems. In physics, coherence is used as a measure of the possible interference of waves. Two waves are coherent when they are capable of forming an interference pattern and storing information. On the basis of the principle of coherent systems created through self-organization, Romijn posits that the constantly changing electric and magnetic fields of the neuronal networks can be seen as a biological quantum-coherence phenomenon. It creates the conditions for complementary systems. This would make the electromagnetic fields, which Romijn believes may be based on "virtual" photons, meaning seemingly or possibly real photons, the carriers or the product of nonlocal consciousness. There is general agreement about the (extremely short) existence of virtual particles, being in a constant process of creation and annihilation. By viewing the electromagnetic fields as a biological quantum phenomenon, Romijn avoids the criticism that the brain is a macroscopic, warm system that naturally causes decoherence (the leak of information) and therefore rules out quantum processes. Given the nonperiodic (unpredictable) nature of consciousness, he proposes a complementary theory with a hitchhiking consciousness capable of translating the physical periodicity (regular recurrence) of dead matter into the nonperiodic processes of living matter in nonlocal space.

The process shows a certain analogy with the double-slit experiment, in which as soon as the intensity of the light dwindles from a massive bombardment to the transmission of individual photons there is a shift from an electromagnetic wave to a probability wave. In the case of a single photon, no electromagnetic wave can be measured, but the (immeasurable) probability wave is used to statistically predict where the photon will hit the photographic plate. Perhaps we could apply this to the brain, with brain activity measured through the registration of the electromagnetic field (EEG). In the event of a cardiac arrest this electromagnetic activity will slow to individual pulses with extremely low electromagnetic energy so that these minimal energy packets (pulses) come to resemble individual photons. These minimal energy packets must then be described with the probability waves from quantum physics instead of the electromagnetic waves from classical physics. When the electromagnetic activity can no longer be measured, it does not mean that there are no more probability waves. In fact, this is where the probability wave becomes a useful descriptor. In theory, the complete loss of brain function is still accompanied by (immeasurable) probability waves. Any potential influence on the minimal processes occurring in the brain at that moment cannot be ruled out (the neurons' pilot-light state). NDE studies suggest that during the loss of all measurable brain function people continue to experience nonlocal consciousness; this nonlocal consciousness is theoretically based on probability waves.

In our opinion there is a strong preference for the model of (reciprocal) information transfer between nonlocal consciousness and the brain via quantum spin coherence, with a possible role for (virtual) photons. A recent article in *Nature* provided evidence of quantum coherence in photo synthesis in living systems, whereby solar energy (photons) was converted into chemical energy by wavelike energy transfer through quantum coherence of coherent electronic oscillations in both" the donor and acceptor molecules. This link between electronic and molecular oscillating states is a result of resonance triggered by the superposition of interference patterns of wave functions of energy (photons). In other words, what we are seeing here is nonlocal energy transfer in living systems on the basis of the quantum coherence of photons, which is akin to the process of nonlocal information transfer in the brain through (virtual?) photons.

Photons (waves or particles) are intrinsic quantum objects and natural long-distance carriers of information both in classical communication via radio, TV, mobile phones, and wireless Internet and in quantum communication. In *Science* and *Nature* recently the results were published of research carried out under laboratory conditions that proved information transfer between matter and light through electron spin and nuclear spin resonance on the basis of nonlocal quantum entanglement. This form of information transfer between light and matter is comparable to reciprocal information transfer between nonlocal consciousness and the brain via the model of nuclear spin correlation or nuclear spin coherence.

Recent studies among volunteers have found strong indications of a nonlocal therapeutic effect of certain drugs such as morphine, when the substance was placed between a pulsating magnetic source and the brain. The subjective therapeutic effect in these volunteers was identical to the effect of receiving this drug directly into the body. And the same subjective therapeutic effect was achieved when the subjects drink water that had been exposed to a pulsating magnetic source, to Igser light, microwaves, or even to a flashlight, with the drug placed i between the photon source and the water. The authors ascribe this Empirically proven positive effect to quantum entanglement between nuclear spin and/or electron spin in the water and nuclear spin and/or electron spin in the brain. The nonlocal information transfer is made possible by, respectively, the magnetic, laser, or flashlight source or the microwaves.

In conclusion, these three possible models of an interface between nonlocal consciousness and the brain will have to be elaborated though future research because the questions continue to outnumber the answers. As mentioned, nonlocal and reciprocal information exchange etween consciousness and the brain will never be fully knowable or Verifiable, rendering any theories on the subject by definition difficult ;-p prove or disprove. Perhaps a combination of data from empirical nd theoretical scientific research could contribute to more definitive answers.

On the strength of the prospective studies of near-death experience and recent data from neurophysiological research and concepts from quantum theory, we strongly believe that consciousness cannot be localized in any particular place—not even in the brain. It is nonlocal (that is, everywhere) in the form of probability waves. For this reason it cannot be demonstrated or measured in the physical world. There is, independent of the body, a continuity of consciousness that is intrinsically connected to or entangled in nonlocal space, though not identical to this space. The different aspects of consciousness are all nonlocal and accessible, although there is probably some kind of hierarchy. The essence or foundation of consciousness (protoconsciousness) probably lies in the vacuum or plenum of the universe, from where it has a nonlocal connection with consciousness in nonlocal space (pariproto-psychism). In this view, the vacuum is the source both of the physical world and of consciousness. Perhaps nonlocal space could be called the absolute or true vacuum because the vacuum and nonlocal space are either identical or nonlocally connected and therefore indistinguishable. Everything is a form of space. Consciousness encompasses nonlocal space, and both my consciousness and yours encompass all space. In fact, each part of our consciousness encompasses all space because each part of infinite is infinite itself. This is exactly what the concept of nonlocality means.

Nonlocal consciousness is the source of our waking consciousness. The two are complementary aspects of consciousness. Under normal, everyday circumstances people experience waking consciousness (the "particle" aspect), which is just one small part of overall and endless nonlocal consciousness (the "wave function" aspect). During life people perceive with the senses while the brain functions as interface. Under abnormal circumstances, people can experience the endless aspect of nonlocal consciousness independent of the body, which is called the continuity of consciousness, and perceive directly via consciousness in space. This is known as a near-death experience. DMT from the pineal gland, of which the release seems to be triggered or stimulated by events in our consciousness, could play a key role in establishing and disrupting the interface between the brain and nonlocal consciousness. As mentioned, this interface may be based on quantum spin coherence (nuclear spin resonance). Nonlocal consciousness is endless, just as each part of consciousness is endless. But our body is not endless. Every day, fifty billion cells are broken down and regenerated in our body. And yet we experience our body as continuous. Where does the continuity of the constantly changing body come from? How can we explain long-term memory if the molecular composition of the neurons' cell membrane is completely renewed every two weeks? And how can we have a long-term memory if the millions of synapses in the brain undergo a process of constant adaptation (neuroplasticity) ? The next chapter will consider these questions in more detail.

MODIFIED SCHRODINGER EQUATION

When M. Planck made the first quantum discovery he noted an interesting fact. The speed of light, Newton's gravity constant and Planck's constant clearly reflect fundamental properties of the world. From them it is possible to derive the characteristic mass M_P, length L_P and time T_P with approximate values

$L_P = 10^{-35}$ m
$T_P = 10^{-43}$ s
$M_P = 10^{-5}$ g.

The constants L_P. T_P *and* M_P. describe Planck Epoch. The enormous efforts of the physicists, mathematicians and philosophers investigate the Planck Epoch. In the subsequent we argue that the source of the "hard "consciousness phenomena are routed in Planck Epoch [1].

To start with we derive modified Schrodinger equation from the study of the thermal phenomena. The thermal history of the system (brain,Universe) can be described by the generalized Fourier equation

$$q(t) = -\int_{-\infty}^{t} \underbrace{K(t-t')}_{\text{thermal history}} \underbrace{\nabla T(t')dt'}_{\text{diffusion}}. \tag{23.1}$$

In Eq. (23.1) $q(t)$ is the density of the energy flux, T is the temperature of the system and $K(t-t')$ is the thermal memory of the system

$$K(t-t') = \frac{K}{\tau} \exp\left[-\frac{(t-t')}{\tau}\right], \tag{23.2}$$

where K is constant, and τ denotes the relaxation time.

As was shown in [1]

$$K(t-t') = \begin{cases} K\delta(t-t') & \text{diffusion} \\ K = \text{constant} & \text{wave} \\ \frac{K}{\tau}\exp\left[-\frac{(t-t')}{\tau}\right] & \text{damped wave or hyperbolic diffusion.} \end{cases}$$

The damped wave or hyperbolic diffusion equation can be written as:

$$\frac{\partial^2 T}{\partial t^2} + \frac{1}{\tau}\frac{\partial T}{\partial t} = \frac{D_T}{\tau}\nabla^2 T. \tag{23.3}$$

For $\tau \to 0$, Eq. (3.3) is the Fourier thermal equation

$$\frac{\partial T}{\partial t} = D_T \nabla^2 T \tag{23.4}$$

and D_T is the thermal diffusion coefficient. The systems with very short relaxation time have very short memory. On the other hand for $\tau \to \infty$ Eq. (23.3) has the form of the thermal wave (undamped) equation, or *ballistic* thermal equation. In the solid state physics the *ballistic* phonons or electrons are those for which $\tau \to \infty$. The experiments with *ballistic* phonons or electrons demonstrate the existence of the *wave motion* on the lattice scale or on the electron gas scale.

$$\frac{\partial^2 T}{\partial t^2} = \frac{D_T}{\tau} \nabla^2 T. \tag{23.5}$$

For the systems with very long memory Eq. (23.3) is time symmetric equation with no arrow of time, for the Eq. (23.5) does not change the shape when $t \to -t$.

In Eq. (23.3) we define:

$$\upsilon = \left(\frac{D_T}{\tau} \right), \tag{23.6}$$

velocity of thermal wave propagation and

$$\lambda = \upsilon \tau, \tag{23.7}$$

where λ is the mean free path of the heat carriers. With formula (23.6) equation (23.3) can be written as

$$\frac{1}{\upsilon^2} \frac{\partial^2 T}{\partial t^2} + \frac{1}{\tau \upsilon^2} \frac{\partial T}{\partial t} = \nabla^2 T. \tag{23.8}$$

From the mathematical point of view equation:

$$\frac{1}{\upsilon^2} \frac{\partial^2 T}{\partial t^2} + \frac{1}{D} \frac{\partial T}{\partial t} = \nabla^2 T$$

is the hyperbolic partial differential equation (PDE). On the other hand Fourier equation

$$\frac{1}{D} \frac{\partial T}{\partial t} = \nabla^2 T \tag{23.9}$$

and Schrödinger equation

$$i\hbar \frac{\partial \Psi}{\partial t} = -\frac{\hbar^2}{2m} \nabla^2 \Psi \tag{23.10}$$

are the parabolic equations. Formally with substitutions

$$t \leftrightarrow it, \ \Psi \leftrightarrow T \tag{23.11}$$

Fourier equation (23.9) can be written as

$$i\hbar\frac{\partial\Psi}{\partial t}=-D\hbar\nabla^{2}\Psi \tag{23.12}$$

and by comparison with Schrödinger equation one obtains

$$D_{T}\hbar=\frac{\hbar^{2}}{2m} \tag{23.13}$$

and

$$D_{T}=\frac{\hbar}{2m}. \tag{23.14}$$

Considering that $D_{T}=\tau\upsilon^{2}$ (23.6) we obtain from (23.14)

$$\tau=\frac{\hbar}{2m\upsilon_{h}^{2}}. \tag{23.15}$$

Formula (23.15) describes the relaxation time for quantum thermal processes. Starting with Schrödinger equation for particle with mass m in potential V:

$$i\hbar\frac{\partial\Psi}{\partial t}=-\frac{\hbar^{2}}{2m}\nabla^{2}\Psi+V\Psi \tag{23.16}$$

and performing the substitution (23.11) one obtains

$$\hbar\frac{\partial T}{\partial t}=\frac{\hbar^{2}}{2m}\nabla^{2}T-VT \tag{23.17}$$

$$\frac{\partial T}{\partial t}=\frac{\hbar}{2m}\nabla^{2}T-\frac{V}{\hbar}T. \tag{23.18}$$

Equation (23.18) is Fourier equation (parabolic PDE) for $\tau = 0$. For $\tau \neq 0$ we obtain

$$\tau\frac{\partial^{2}T}{\partial t^{2}}+\frac{\partial T}{\partial t}+\frac{V}{\hbar}T=\frac{\hbar}{2m}\nabla^{2}T, \tag{23.19}$$

$$\tau=\frac{\hbar}{2m\upsilon^{2}}$$

Where the relaxation time τ is the real constant. Considering (2.19) we obtain

$$\frac{1}{\upsilon^2}\frac{\partial^2 T}{\partial t^2}+\frac{2m}{\hbar}\frac{\partial T}{\partial t}+\frac{2Vm}{\hbar^2}T=\nabla^2 T. \tag{23.20}$$

With the substitution (23.11) equation (23.19) can be written as

$$i\hbar\frac{\partial\Psi}{\partial t}=V\Psi-\frac{\hbar^2}{2m}\nabla^2\Psi-\tau\hbar\frac{\partial^2\Psi}{\partial t^2}. \tag{23.21}$$

The new term, relaxation term

$$\tau\hbar\frac{\partial^2\Psi}{\partial t^2} \tag{23.22}$$

describes the interaction of the particle with mass m with space-time. The relaxation time τ can be calculated as:

$$\tau^{-1}=\left(\tau_{e-p}^{-1}+\ldots+\tau_{Planck}^{-1}\right), \tag{23.23}$$

where, for example $\tau_{e\text{-}p}$ denotes the scattering of the particle m on the electron-positron pair ($\tau_{e-p}\sim 10^{-17}$s) and the shortest relaxation time τ_{Planck} is the Planck time ($\tau_{Planck}\sim 10^{-43}$s).

From equation (23.23) we conclude that $\tau\approx\tau_{Planck}$ and equation (23.21) can be written as

$$i\hbar\frac{\partial\Psi}{\partial t}=V\Psi-\frac{\hbar^2}{2m}\nabla^2\Psi-\tau_{Planck}\hbar\frac{\partial^2\Psi}{\partial t^2}, \tag{23.24}$$

where

$$\tau_{Planck}=\frac{1}{2}\left(\frac{\hbar G}{c^5}\right)^{\frac{1}{2}}=\frac{\hbar}{2M_p c^2}. \tag{23.25}$$

In formula (23.25) M_p is the mass Planck. Considering Eq. (2.25), Eq. (2.24) can be written as

$$i\hbar\frac{\partial\Psi}{\partial t}=-\frac{\hbar^2}{2m}\nabla^2\Psi+V\Psi-\frac{\hbar^2}{2M_p}\nabla^2\Psi+\frac{\hbar^2}{2M_p}\nabla^2\Psi-\frac{\hbar^2}{2M_p c^2}\frac{\partial^2\Psi}{\partial t^2}. \tag{23.26}$$

The last two terms in Eq. (23.26) can be defined as the *Bohmian* pilot wave

$$\frac{\hbar^2}{2M_p}\nabla^2\Psi - \frac{\hbar^2}{2M_p c^2}\frac{\partial^2\Psi}{\partial t^2} = 0, \tag{23.27}$$

i.e.

$$\nabla^2\Psi - \frac{1}{c^2}\frac{\partial^2\Psi}{\partial t^2} = 0. \tag{23.28}$$

It is interesting to observe that pilot wave Ψ does not depend on the mass of the particle. With postulate (23.28) we obtain from equation (23.26)

$$i\hbar\frac{\partial\Psi}{\partial t} = -\frac{\hbar^2}{2m}\nabla^2\Psi + V\Psi - \frac{\hbar^2}{2M_p}\nabla^2\Psi \tag{23.29}$$

and simultaneously

$$\frac{\hbar^2}{2M_p}\nabla^2\Psi - \frac{\hbar^2}{2M_p c^2}\frac{\partial^2\Psi}{\partial t^2} = 0. \tag{23.30}$$

In the operator form Eq. (23.21) can be written as

$$\hat{E} = \frac{\hat{p}^2}{2m} + \frac{1}{2M_p c^2}\hat{E}^2, \tag{23.31}$$

where $\hat{E}$ and $\hat{p}$ denote the operators for energy and momentum of the particle with mass m. Equation (23.31) is the new dispersion relation for quantum particle with mass m. From Eq. (23.21) one can concludes that Schrödinger quantum mechanics is valid for particles with mass $m \ll M_P$. But pilot wave exists independent of the mass of the particles.

For particles with mass $m \ll M_P$ Eq. (23.29) has the form

$$i\hbar\frac{\partial\Psi}{\partial t} = -\frac{\hbar^2}{2m}\nabla^2\Psi + V\Psi. \tag{23.32}$$

In the case when $m \approx M_p$ Eq. (2.29) can be written as

$$i\hbar\frac{\partial\Psi}{\partial t}=-\frac{\hbar^2}{2M_p}\nabla^2\Psi+V\Psi, \tag{23.33}$$

but considering Eq. (23.30) one obtains

$$i\hbar\frac{\partial\Psi}{\partial t}=-\frac{\hbar^2}{2M_pc^2}\frac{\partial^2\Psi}{\partial t^2}+V\Psi \tag{23.34}$$

or

$$\frac{\hbar^2}{2M_pc^2}\frac{\partial^2\Psi}{\partial t^2}+i\hbar\frac{\partial\Psi}{\partial t}-V\Psi=0. \tag{23.35}$$

We look for the solution of Eq. (23.35) in the form

$$\Psi(x,t)=e^{-i\omega}u(x). \tag{23.36}$$

After substitution formula (23.36) to Eq. (23.35) we obtain

$$\frac{\hbar^2}{2M_pc^2}\omega^2-\omega\hbar+V(x)=0 \tag{23.37}$$

with the solution

$$\omega_1=-\frac{M_pc^2+M_pc^2\sqrt{1-\frac{2V}{M_pc^2}}}{\hbar}$$
$$\omega_2=\frac{M_pc^2-M_pc^2\sqrt{1-\frac{2V}{M_pc^2}}}{\hbar} \tag{23.38}$$

for $\frac{M_pc^2}{2}>V$ and

$$\omega_1 = \frac{M_p c^2 + iM_p c^2 \sqrt{\frac{2V}{M_p c^2} - 1}}{\hbar}$$

$$\omega_2 = \frac{M_p c^2 - iM_p c^2 \sqrt{\frac{2V}{M_p c^2} - 1}}{\hbar} \tag{23.39}$$

for $\frac{M_p c^2}{2} < V$.

Both formulae (23.38) and (23.39) describe the "string "oscillation, formula (23.27) damped 2oscillation and formula (23.28) over damped "string "oscillation.

From elementary particles physics we know that the internal energy $M_P c^2$ is the maximum energy per particle in the Universe (for elementary particles (Figure 1). In that case we argue that the first solution (23.38) is the valid solution.

For $\frac{M_p c^2}{2} < V$ we obtain

$$\omega_1 = \frac{2M_p c^2}{\hbar}$$

$$\omega_2 = \frac{V}{\hbar} \tag{23.40}$$

The angular frequency ω_1 represent the so called Planck frequency $\omega_1 = \tau_P^{-1}$ and ω_2 is the frequency for the brain waves.

Figure 2 presents the ω_1 as the function of the ratio

$$\frac{2V}{M_p c^2} \tag{23.41}$$

As can be seen from Figure 2. for the potential energy $V \approx 10^{-15} eV$ angular frequency of the brain waves is of the order of *10 Hz*

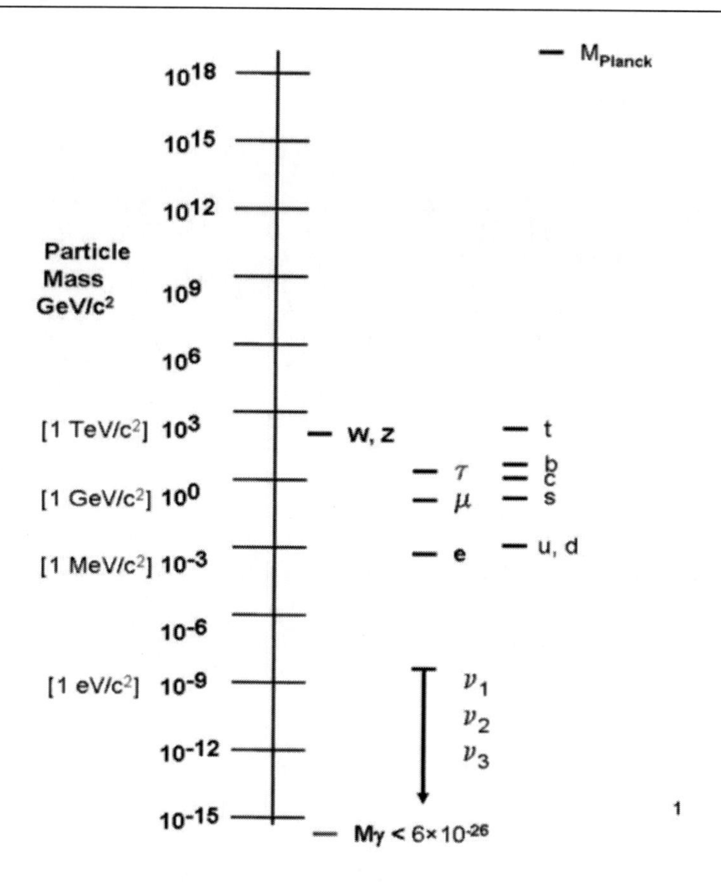

Figure 1. The energy scale for elementary particle physics.

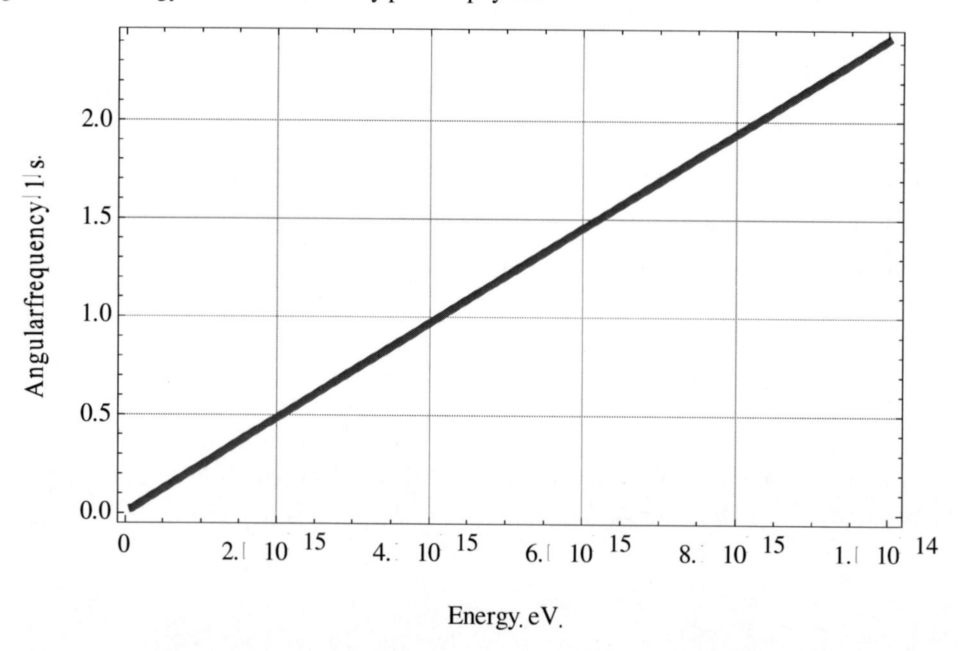

Figure 2. Angular frequency as the function of energy.

REFERENCE

[1] J. Marciak-Kozlowska, M. Kozlowski, From femto to attoscience and beyond, Nova science publishers, 2009.

Chapter 24

HEISENBERG UNCERTAINTY AND THE HUMAN BRAIN

The issue of observation in QM is central, in the sense that objective reality cannot be disentangled from the act of observation, as the Copenhagen Interpretation (CI) nearly states In the words of John A. Wheeler, we live in an observer-participatory Universe. The vast majority of today's practicing physicists follow CI's practical prescriptions for quantum phenomena, while still clinging to classical beliefs in observer-independent local, external reality). There is a critical gap between practice and underlying theory. In his Nobel Prize speech of 1932, Werner Heisenberg concluded that the atom "has no immediate and direct physical properties at all." If the universe's basic building block isn't physical, then the same must hold true in some way for the whole. The universe was doing a vanishing act in Heisenberg's day, and it certainly hasn't become more solid since. This discrepancy between practice and theory must be confronted, because the consequences for the nature of reality are far-reaching An impressive body of evidence has been building to suggest that reality is non-local and undivided. Non-locality is already a basic fact of nature, first implied by the Einstein-Podolsky-Rosen thought experiment despite the original intent to refute it, and later explicitly formulated in Bell's Theorem. Moreover, this is a reality where the mindful acts of observation play a crucial role at every level. Heisenberg again: "The atoms or elementary particles themselves ... form a world of potentialities or possibilities rather than one of things or facts." He was led to a radical conclusion that underlies our own view in this paper: "What we observe is not nature itself, but nature exposed to our method of questioning." Reality, it seems, shifts according to the observer's conscious intent. There is no doubt that the original CI was subjective Quantum theory is not about the nature of reality, even though quantum physicists act as if that is the case. To escape philosophical complications, the original CI was pragmatic: it concerned itself with the epistemology of quantum world (how we experience quantum phenomena), leaving aside ontological questions about the ultimate nature of reality. The practical bent of CI should be kept in mind, particularly as there is a tendency on the part of many good physicists to slip back into issues that cannot be tested and therefore run counter to the basic tenets of scientific methodology.

HEISENBERG'S UNCERTAINTY

*Q*uantum physics, as exemplified by the Copenhagen school (Bohr,; Heisenberg,, also makes assumptions about the nature of reality as related to an observer, the "knower" who is conceptualized as a singularity. Because the physical world is relative to being known by a "knower" (the observing consciousness), then the "knower" can influence the nature of the reality which is being observed. In consequence, what is known versus what is not known becomes relatively imprecise For example, as expressed by the Heisenberg uncertainty principle the more precisely one physical property is known the more unknowable become other properties, whose measurements become correspondingly imprecise. The more precisely one property is known, the less precisely the other can be known and this is true at the molecular and atomic levels of reality. Therefore it is impossible to precisely determine, simultaneously, for example, both the position and velocity of an electron. However, we must ask: if knowing A, makes B unknowable, and if knowing B makes A unknowable, wouldn't this imply that both A and B, are in fact unknowable? If both A and B are manifestations of the processing of "knowing," and if observing and measuring can change the properties of A or B, then perhaps both A and B are in fact properties of knowing, properties of the observing consciousness, and not properties of A or B. In quantum physics, nature and reality are represented by the quantum state. The electromagnetic field of the quantum state is the fundamental entity, the continuum that constitutes the basic oneness and unity of all things. The physical nature of this state can be "known" by assigning it mathematical properties. Therefore, abstractions, i.e., numbers, become representational of a hypothetical physical state. Because these are abstractions, the physical state is also an abstraction and does not possess the material consistency, continuity, and hard, tangible, physical substance as is assumed by Classical (Newtonian) physics. Instead, reality, the physical world, is created by the process of observing, measuring, and knowing Consider an elementary particle, once this positional value is assigned, knowledge of momentum, trajectory, speed, and so on, is lost and becomes "uncertain." The particle's momentum is left uncertain by an amount inversely proportional to the accuracy of the position measurement which is determined by values assigned by the observing consciousness. Therefore, the nature of reality, and the uncertainty principle is directly affected by the observer and the process of observing and knowing The act of knowing creates a knot in the quantum state; described as a "'collapse of the wave function;" a knot of energy that is a kind of blemish in the continuum of the quantum field. This quantum knot bunches up at the point of observation, at the assigned value of measurement.

The process of knowing, makes reality, and the quantum state, discontinuous. "The discontinuous change in the probability function takes place with the act of registration... in the mind of the observer" (Heisenberg)Reality, therefore, is a manifestation of alterations in the patterns of activity within the electromagnetic field which are perceived as discontinuous. The perception of a structural unit of information is not just perceived, but is inserted into the quantum state which causes the reduction of the wave-packet and the collapse of the wave function. Knowing and not knowing, are the result of interactions between the mind and concentrations of energy that emerge and disappear back into the electromagnetic quantum field. However, if reality is created by the observing consciousness, and can be made discontinuous, does this leave open the possibility of a reality behind the reality? Might there be multiple realities? And if consciousness and the observer and the quantum state is not a singularity, could each of these multiple realities also be manifestations of a multiplicity of minds? Heinsenberg recognized this possibility of hidden realities, and therefore proposed that the reality that exists beyond or outside the quantum state could better understood when considered in terms of "potential" reality and "actual" realities. Therefore, although the quantum state does not have the ontological character of an "actual" thing, it has a "potential" reality; an objective tendency to become actual at some point in the future, or to have become actual at some it in the past. Therefore, it could be said that the subatomic particles which make up reality, or quantum state, do not really exist, except as probabilities. These "subatomic" particles have probable existences and display tendencies to assume certain patterns of activity that we perceive as shape and form. Yet, they may also begin - -play a different pattern of activity such that being can become nonbeing and something else altogether. The conception of a deterministic reality is therefore subjugated to mathematical probabilities and potentiality which is relative to the mind of a knower which effects that reality as it unfolds, evolves, and is observed. That is, the mental act of perceiving a non-localized unit of structural information, injects that mental event into the quantum state of the universe, causing "the collapse of the wave function" and creating a bunching up, a tangle and discontinuous knot in the continuity of the quantum state. Heisenberg), cautioned, however, that the observer is not the creator of reality: "The introduction of the observer must not be misunderstood to imply that some kind of subjective features are to be brought into the description of nature. The observer has, rather, only the function of registering decisions, i.e., processes in space and time, and it does not matter whether the observer is an apparatus or a human being; but the registration, i.e., the transition from the "possible" to the "actual," is absolutely necessary here and cannot be omitted from the interpretation of quantum theory."

Shape and form are a function of our perception of dynamic interactions within the continuum which is the quantum state. What we perceive as mass (shape, form, length, weight) are dynamic patterns of energy which we selectively attend to and then perceive as stable and static, creating discontinuity within the continuity of the quantum state. Therefore, what we are perceiving and knowing, are only fragments of the continuum.

However, we can only perceive what our senses can detect, and what we detect as form and shape is really a mass of frenzied subatomic electromagnetic activity that is amenable to detection by our senses and which may be known by a knowing mind. It is the perception of certain aspects of these oscillating patterns of continuous evolving activity, which give rise to the impressions of shape and form, and thus discontinuity, as experienced within the mind. This energy that makes up the object of our perceptions, is therefore but an aspect of the electromagnetic continuum which has assumed a specific pattern during the process of being sensed and processed by those regions of the brain and mind; best equipped to process this information. Perceived reality, therefore, becomes a manifestation of mind.

However, if the mind is not a singularity, and if we possessed additional senses or an increased sensory channel capacity, we would perceive yet other patterns ar.: other realities which would be known by those features of the mind best attune: to them. If the mind is not a singularity but a multiplicity, this means that both - and B, may be known simultaneously.

$$\Delta E \Delta t = \hbar \quad \Delta E = 0 \rightarrow \Delta t = \infty$$

MODEL

In order to put forward the classical theory of the brain waves we quantize the brain wave field. In the model presented herewe assume (i) the brain is the thermal source in local equilibrium with temperature T.(ii) The spectrum of the brain waves is quantized according to formula

$$E = \hbar\omega \tag{24.1}$$

where E is the photon energy in eV, $\hbar$=Planck constant, $\omega = 2\pi\nu, \nu$ -is the frequency in Hz. (iii). The number of photons emitted by brain is proportional to the $(\text{amplitude})^2$ as for classical waves. The energies of the photons are the maximum values of energies of waves For the emission of black body brain waves we propose the well know formula for the black body radiation. In thermodynamics we consider Planck type formula for probability P (E) dE for the emission of the particle (photons as well as particles with m≠0) with energy (E,E+dE) by the source with temperature T is equal to:

$$P(E)dE = BE^2\, e^{(-E/kT)}\, dE \tag{24.2}$$

where B= normalization constant, E=total energy of the particle, k = Boltzmann constant=1.3 x 10^{-23} J K^{-1}. K is for Kelvin degree. However in many applications in nuclear and elementary particles physics kT is recalculated in units of energy. To that aim we note that for 1K, kT is equal k1K = K x 1. 3 10^{-23} J x K^{-1}= 1.3 10^{-23} Joule or kT for 1K is equivalent to 1.3 10^{-23} Joule= 1.3 10^{-23} /(1.6 10^{-19}) eV = 0.8 10^{-4} eV. Eventually we obtain 1K= 0.8 10^{-4} eV, and 1eV= 1.2 10^4 K

$$\frac{dN}{dE} = BE_{\max}^2 e^{(-\frac{E_{\max}}{T})} \tag{24.3}$$

where, B is the normalization constant, T is the temperature of the brain thermal source in eV. The function $\frac{dN}{dE}$ describes the energy spectrum of the emitted brain photons.

In Figure 1 the calculated energy spectrum, formula (24.3) is presented. We present the result of the comparison of the calculated and observed spectra of the brain waves. The calculated spectra are normalized to the maximum of the measured spectra. The calculated spectrum is for temperature of brain source T= 0.8 10^{-14} eV. The obtained temperature is the temperature for the brain source in the thermal equilibrium. The source is thermally isolated (adiabatic well). However in very exceptional cases the spectrum is changed – by the tunneling to the quantum potential well. The temperature 1 eV $\cong$ 10 ^{4} K then brain wave thermal spectra T=0.8 10^{-14} eV= 0.8 10^{-10} K.

In Figure 2 we present the calculation of the energy spectrum for the Cosmic Background Radiation. Tke formula (2) was used for the model calculation. The normalized theoretical spectrum describe very well the observed CBR. The calculated temperature T=2.53K, which is in excellent agreement with experimentally verified values

It must be stressed that we abandon the idea that every physical object is either a wave or a particle. Neither it is possible to say that particles "become" waves in the quantum domain and conversely that waves are "transformed "into particles. It is therefore necessary to acknowledge that we have here a different kind of an entity, one that is specifically quantum. For this reason Levy-Leblond and Balibar developed the name *quanton,* (Levy-Leblond),. Following that idea the human brain emits *quantons* with energies $E=\hbar\omega$ formula (24.3). The brain *quantons* are the quantum objects that follows all quantum laws: tunneling,, the superposition and Heisenberg uncertainty rule. For the wave length of the *quantons* is of the order of Earth radius the quantum nature of the brain will be manifested in the Earth scale.

The formula for Heisenberg uncertainty formula can written as

$$\Delta E \Delta t = \hbar \tag{24.4}$$

Where E is the characteristic energy of the system and t is the characteristic time. From formula (3) we can calculate the characteristic times for energy of the sources. In Table 1 the result of the calculations for characteristic times, formula (3) are presented. According to Libet theory, the characteristic time for the nerve brain response is of the order of 1.4 s.

In history of the Universe the characteristic time for appearance of all interactions: gravity, electromagnetic, electroweak, and strong is of the order of 10^{-11} s.

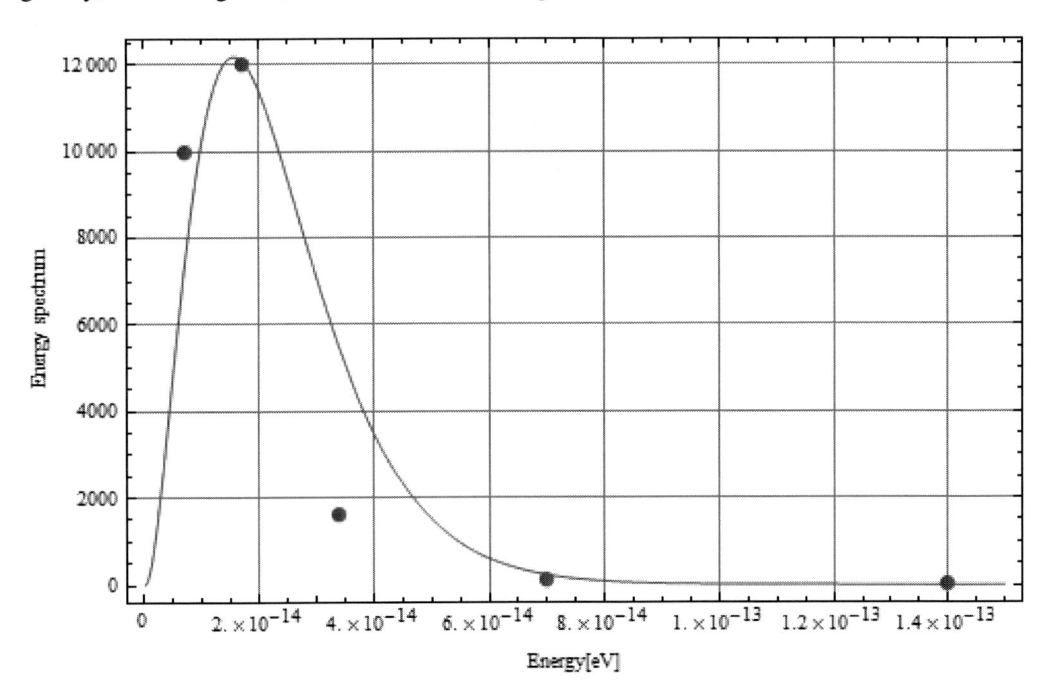

Figure 1. Model calculations for energy spectra of brain photons. The temperature of the source, T= 7.8 10^{-11} K.

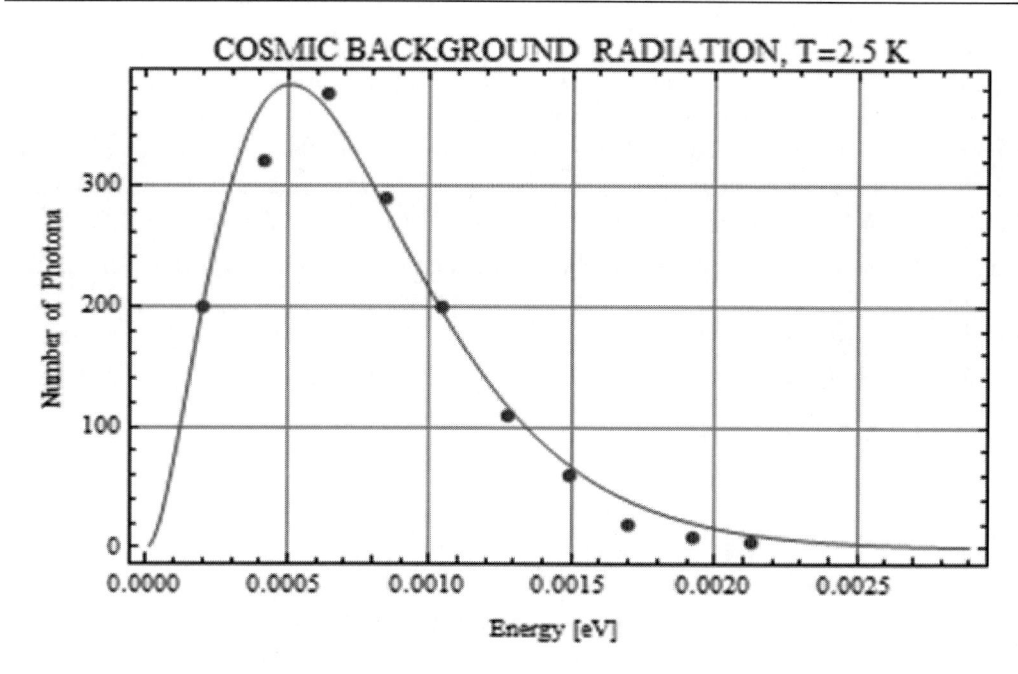

Figure 2. Model calculations for energy spectra of Cosmic Background Radiation Temperature of the source T= 2.35 K.

Table 1. Characteristic times for Human Brain and Universe

Source	Energy [eV] (This paper)	Characteristic time, [s] (Heisenberg inequality)
Human Brain	$7.8\ 10^{-15}$	0.45
Cosmic Background Radiation	$2.5\ 10^{-4}$	$3.1\ 10^{-11}$

Table 2. Hypothesis

Source	Energy[eV] (This paper)	Time[s] (This paper)	Comparison to: Libet and Perkins
Brain	$7.8\ 10^{-15}$	0.45	Libet, 1.4 s
Universe	$2.5\ 10^{-4}$	$3.1\ 10^{-11}$	Perkins, 10^{-11} s

Chapter 25

HEAVISIDE QUANTONS AS THE CARRIERS OF THE ESP PHENOMENA

In this paragraph we investigate the EPS phenomena with the help of the contemporary elementary particle physics. We assume that ESP phenomena are rooted in subnuclear physics. First of all supported by the results of the Oliver Heaviside we conclude that special relativity is not in opposition to the existence of the particles with *finite mass and velocities greater than light velocity*. The spark of ESP phenomenon in "source" subject is created by the emission of the new particle- antiparticle *Heaviside quantons*, which consists of *Heaviside particles* with mass of the order of 10^{-15} eV, which propagate with velocity greater than the light velocity. The recombination of the Heaviside pair generates the additional Hydrogen atom in the brain medium of the receiving subject

1. ESP DATA

The type of data which can be acquired in ESP phenomena can be summarized as:

1. The overall ambience of the scene is accurately perceived.
2. Certain details are accurately identified; others are misconstrued or totally ignored.
3. A feature which is impressive to the agent is not necessarily so to the percipient, and vice versa.
4. The composition of the scene may be distorted by errors in scale, relative positions of key objects, or total right- left inversions.
5. The aesthetic aspects, such as colors, general shapes, degree of activity, noise level, climate, and other ambient features tend to be more accurately perceived than more analytical details such as number, size, or relative positions.
6. The perception is not necessarily centered on the defined target, and may even provide accurate information on adjacent areas external to the target, unnoticed by the agent.
7. *The fidelity of the perception seems to be independent of the remoteness of the target, up to distances of several thousand miles.*

8. *The time of the perception effort need not coincide with the time the agent is at the target. Perceptions obtained several hours, or even days, prior to the agent's visit to the target, or even prior to selection of the target, display at least as high a yield as those performed in real time.*

The philosophical and practical implications of items 7 and 8 are clearly substantial. If the data are valid, the most parsimonious explications would require access of the percipient's consciousness to other portions of the space-time grid than that in which it is currently immersed, or that it can reach by normal processes of communication or memory. These same items also seriously delimit the potential physical mechanisms for such access.

In this paper we develop the new model for ESP phenomena based on pre- Einseinian relativity theory. The model we developed takes into account that relativity theory not precludes existence of the particles with finite mass and velocity greater than light velocity in vacuum. Oliver Heaviside formulated the theory of superluminal particles in XIX century. (Heaviside,1885). Following the Heaviside`s theory we postulate the new charged particle with mass m= 10^{-15} eV, Heaviside particle which is responsible for the ESP phenomena. The pilot Heaviside particle emits Heaviside electric field (velocity=c) in very narrow angle θ,

$$Tan(\theta) = \frac{1}{\sqrt{\frac{u^2}{c^2} - 1}}$$

In a sense the subject (emitter) “sees “the subject (receiver)

2. Physics of the ESP

Action at Distance

First of all we will discuss the basis for ESP given by physics. Let`s start with action at distance (points 7 and 8) Ampere was, indeed, a man of genius, but even after his work on interacting electrical currents there were still great, puzzles to be addressed. The major difficulty was that of the ***nature*** of the interaction between currents (or electrically charged bodies). This interaction involved, obviously, forces bodies, but these forces were not produced by anything obviously pushing or pulling on bodies. It was ***action-at-a-distance.*** Action-at-a-distance was nothing new in physics Ampere's time, as Newton himself had been faced with the same concern in his theory of masses interacting gravitationally and instantaneously across the empty vastness of spaces. Newton was not happy about action-at-a-distance, and indeed his gravitational theory attacked by many who claimed it was a reversion to "explaining" Nature by invoking powers. Unable to suggest anything else in place of it he contented himself with his famous passage (Newton, 1693):

That Gravity should be innate, inherent and essential to Matter, so that one body may act upon another at a Distance thro' a *Vacuum,* without the Mediation of anything else, by and through which their Action and Force may be conveyed from one to another, is to me so great an Absurdity, that I believe no Man who has in philosophical Matters a competent Faculty of thinking, can ever fall into it. Gravity must be caused by an Agent acting constantly according to certain Laws; but whether this Agent be material or immaterial.

The Aether

The nature of all forces known to Newton and his contemporaries by direct. ***t~*** experience seemed always to be that of contact, i.e., a push or a pull by the intimate mectu interaction of one thing (via a rope, or a stick, or one's hand, etc.) with another. Gravitation action-at-a-distance (whether instantaneous or not) is most mysterious if acting in a meets way through a vacuum which is truly empty. But suppose that even a vacuum is filled i substance that can transmit forces, a substance something like air but ever so much thin: penetrating, a substance that can slip through all of ponderable matter and fill every noo cranny of the universe. Suppose the universe is embedded in an ocean of this mist called (or *aether*)—then what? Interacting bodies, even though *apparently* separated by the empty gulf of a vacuum. then be imagined as actually still in mechanical communion via stresses and strains ii the ether. So attractive is this idea, in fact, that the ether concept can be traced to ancient at least as far back as Aristotle. The price paid for this imaginative idea, however, was one—for every sort of apparent action-at-a-distance phenomenon it was necessary to corresponding ether until, as Maxwell complained (Niven, 1890), 'Aethers were invented for the swim in, to constitute electric atmosphere and magnetic effluvin, and so on, to sensations from one part of our bodies to another, and so on, till all space had been filled or four times over with aethers."

This aether was thought able to transmit wave motion (from the interference expe Thomas Young it was generally known by 1801 that light is a wave phenomenon), muci gas conducts sound waves. Sound waves, however, are longitudinal waves, with the "waving" back-and- forth along the direction of wave propagation. The initially puzzling fact that light can be polarized was, however, incompatible with longitudinal or "back-and-forth" compression waves. Then Young and Augustin Fresnel, in 1817-18, showed how polarization can be explained by *transversal* waves, with the medium "waving" in a direction *perpendicular* to the direction of the wave propagation. This, in turn, made a gaseous ether unthinkable, as it would be unable to support the shear stresses required by a transversal wave. The aether could not, in fact, be a gas at all, but instead must be an elastic, jellylike solid, a bit of imagery due to William Thomson's old friend, G. G. Stokes. The required mechanical properties of such an ether are fantastic, to say the least. This jelly had to be both thin enough for "the planets to swim in" without any observable retardation or deviation from Newton's laws of motion and rigid enough to propagate waves (light) at a speed of 186,000 miles per second. To imagine such a substance is not easy, yet in 1854 William Thomson wrote, (Thomson, 1884):

That there must be a medium forming a continuous material communication throughout space to the remotest visible body is a fundamental assumption in the undulatory Theory of Light. Whether or not this medium is (as appears to me most probable) a continuation of our own atmosphere, its existence is a fact that cannot be questioned... .

The XIX/XX century position in respect to aether was perhaps put best by Heaviside when he wrote (Heaviside 1893):

As regards the ether, it is useless to sneer at it at this time of day. What substitute for it are we to have? Its principal fault is that it is mysterious. That is because we know so little about it. Then we should find out more. That cannot be done by ignoring it. The properties of air, so far as they are known, had to be found out before they became known.

This passage shows an increase in either Heaviside's optimism or desperation, as earlier in 1885 (Heaviside, 1885) he had written,

Ether is a very wonderful thing. It may exist only in the imaginations of the wise, being invented and endowed with properties to suit their hypotheses; but we cannot do without it.... But admitting the ether to propagate gravity instantaneously, it must have wonderful properties, unlike anything we know.

Beyond Special Relativity

The original version of relativity, the so-called special theory, is actually a law and a rather simple one at that, being not an equation of motion at all but a property of that equation, a symmetry. The most mature form of relativity is a speculative post-Newtonian theory of gravity motivated by this law.[2] Einstein, who discovered early in his career that the public was more interested in the mystical aspects of relativity than the physical ones, encouraged the growth of his image as a seer even though he was not a seer at all but a professional with a razor-sharp mind. However, Einstein's writing is characteristically well-reasoned, direct, and open. He was capable of being wrong, just like the rest of us, but he rarely hid his mistakes in abstruse mathematics. Most physicists aspire to be as clear as Einstein, but few of them succeed.

Symmetry is an important, if often abused, idea in physics. An example of symmetry is roundness. Billiard balls are round, and this allows one to make some predictions about them without knowing exactly what they are made of, for example, that they will roll in straight-line paths across the table when struck with a cue. But roundness does not cause them to move. The underlying laws of motion do that. Roundness is just a special property that sets billiard balls apart from arbitrary rigid bodies and is revealed by the unusual simplicity and regularity of their motion. Symmetry is especially helpful in situations where one does not know the underlying equations of motion and is trying to piece them together from incomplete experimental facts. If, for example, you knew that all billiard balls were round and were trying to guess their equations of motion, you could eliminate certain guesses on the grounds that round things could not possibly do this. Situations of this kind are the rule rather than the exception in subnuclear physics. For this reason there is a tradition in physics of ascribing to symmetries an overriding importance even though they are actually a consequence, or property, of the equations of motion. The symmetry of relativity involves motion. Einstein and other early twentieth-century figures came upon this symmetry through thinking about electricity and magnetism, whose equations had just been worked out by James Maxwell and were rapidly leading to the invention of radio. Rotational symmetry requires the behavior of billiard balls on a round table to appear qualitatively the same regardless of where one stands on the perimeter. Relativistic symmetry requires their behavior to appear the same regardless of how one is moving. This idea is captured brilliantly by the famous Einsteinian thought experiment of a passenger on a train watching another train pass by. Einstein proposed that in the ideal limit—two trains passing each other in the vacuum of space—no measurement could determine which train was stationary and which was moving. That being the case, the equations of electricity and magnetism would have to appear the same on the two trains, and thus the speed of light must also be the same. One then encounters a logical contradiction unless some common ideas about simultaneity and measurement on the two trains are wrong. All of these musings and their fascinating logical implications, including the weight gain acquired by objects moving at high speeds and the equivalence of mass and energy, are now routinely verified in laboratories all over the world, and have passed into history as self-evident truth. The story of Einstein's triumph is so romantic it is easy to forget that relativity was a discovery and not an invention. It was subtly implicit in certain early experimental observations about electricity, and it took bold thinking to synthesize these observations into a coherent whole. But no such boldness would be required today. An unsuspecting experimentalist armed with a modern accelerator would stumble upon the effects of relativity the first day and would probably figure the whole thing out empirically in a month. Relativity is actually not shocking at all. The ostensibly self-evident worldview it supplanted was simply based on incomplete and inaccurate observations. Had all the facts been known, there would have been no controversy and thus nothing for Einstein to prove. The popular view of relativity as a creation of the human mind is wonderfully ennobling but in the end incorrect. Relativity was discovered. Einstein's beautiful arguments notwithstanding, we believe in relativity today not because it ought to be true, but because it is measured to be true. Einstein's theory of gravity, in contrast, was an invention, something not on the verge of being discovered accidentally in the laboratory. It is still controversial and largely beyond the reach of experiment.[5] Its most important prediction is that space itself is dynamic. The equations Einstein proposed to describe gravity are similar to those of an elastic medium, such as a sheet of rubber. Conventional gravitational effects result when this medium is distorted statically by a large mass, such as a star. When the source oscillates rapidly, however, such as when two stars revolve around each other in tight orbit, there is a new effect: outwardly propagating ripples of gravity. Conventional gravity is thus like the dimples under the feet of a water skimmer, and gravitational radiation is like the disturbances generated by the skimmer when it scampers away. There is much indirect evidence that the prediction of gravitational radiation is correct, the strongest being the steadily diminishing orbital period of the famous binary pulsar discovered by Joseph Taylor and Russell Hulse in 1975. There is as yet no direct evidence. Detecting gravitational radiation directly is one of the key goals of modern experimental physics,[7] but most physicists are already persuaded by other evidence that Einstein's theory of gravity is probably correct.

It is ironic that Einstein's most creative work, the general theory of relativity, should boil down to conceptualizing space as a medium when his original premise was that no such medium existed. The idea that space might be a kind of material substance is actually very ancient, going back to Greek Stoics and termed by them ether. Ether was firmly in Maxwell's mind when he invented the description of electromagnet-ism we use today. He imagined electric and magnetic fields to be displacements and flows of ether, and borrowed mathematics from the theory of fluids to describe them. Einstein, in contrast, utterly rejected the idea of ether and inferred from its nonexistence that the equations of electromagnetism had to be relative. But this same thought process led in the end to the very ether he had first rejected, albeit one with some special properties that ordinary elastic matter does not have.

The word "ether" has extremely negative connotations in theoretical physics because of its past association with opposition to relativity. This is unfortunate because, stripped of these connotations, it rather nicely captures the way most physicists actually think about the vacuum. In the early days of relativity the conviction that light must be waves of something ran so strong that Einstein was widely dismissed.[8] Even when Michelson and Morley demonstrated that the earth's orbital motion through the ether could not be detected, opponents argued that the earth must be dragging an envelope of ether along with it because relativity was lunacy and could not possibly be right. The virulence of this opposition eventually had the scandalous consequence of denying relativity a Nobel Prize. (Einstein got one anyway, but for other work.) Relativity actually says nothing about the existence or nonexistence of matter pervading the universe, only that any such matter must have relativistic symmetry.

It turns out that such matter exists. About the time relativity was becoming accepted, studies of radioactivity began showing that the empty vacuum of space had spectroscopic structure similar to that of ordinary quantum solids and fluids. Subsequent studies with large particle accelerators have now led us to understand that space is more like a piece of window glass than ideal Newtonian emptiness. It is filled with "stuff" that is normally transparent but can be made visible by hitting it sufficiently hard to knock out a part. The modern concept of the vacuum of space, confirmed every day by experiment, is a relativistic aether. But we do not call it this because it is taboo. How Einstein came to conclude that space was a medium is a fascinating story. His starting point was the principle of equivalence, the observation that all objects fall under the pull of gravity at the same rate regardless of their mass. This is the effect that causes astronauts in near earth orbit to experience weightlessness. The pull of gravity is not significantly smaller in low orbit than on earth, but the effect of this gravity is simply to make them and their spacecraft fall together around the earth. Einstein inferred from this effect (more precisely from versions of it he imagined in 1905 when there were no astronauts) that the force of gravity was inherently fictitious, since it could always be turned off by allowing the observer and his immediate surroundings to fall freely. The important effect of a nearby massive body such as the earth was not to create gravitational forces but to make free-fall paths converge. Astronauts falling straight down onto the earth (an unfortunate experiment) might at first think they were in deep space, but after a while would notice that objects traveling with them were slowly getting closer. This is because all the nearby free-fall paths are directed toward the center of the earth and eventually meet there. Einstein was struck by the similarity between this effect and the convergence of lines of longitude at the north and south poles. In that case, the tendency of some straight-line paths to converge is a consequence of the curvature of the earth—a medium made out of conventional matter. Then, in a flash of insight that leaves us breathless even today, he guessed that free-fall paths actually are lines of longitude on a higher-dimensional surface, and that gravity occurs because large masses stretch this surface and cause it to curve. He then made a second, masterful guess about the specific relation between mass and curvature known to us today as the Einstein field equations. These respect relativity and thus contain the same paradoxes of simultaneity found in the original version of relativity. For this reason they are more accurately described as a relation between stress-energy and the curvature of four-dimensional space-time. Their prediction that space can ripple in addition to stretching is a consequence of its obeying relativity, a symmetry of motion. It is consistent with our physical intuition, however, since it is basically the same thing as a propagating seismic wave on the surface of the earth generated by an earthquake.

The clash between the philosophy of general relativity and what the theory actually says has never been reconciled by physicists and sometimes gives the subject a Kafkaesque flavor. On the one hand, we have the view, founded in the success of relativity, that space is something fundamentally different from the matter moving in it and thus not understandable through analogy with ordinary things. On the other, we have the obvious similarities between Einsteinian gravity and the dynamic warping of real surfaces, leading us to describe space-time as a fabric. Their curiosity is, however, neither naive nor inappropriate. The closet of general relativity contains a horrible skeleton known as the cosmological constant. This is a correction to the Einstein field equations compatible with relativity and having the physical meaning of a uniform mass density of relativistic ether. Einstein originally set this constant to zero on the grounds that no such effect seemed to exist. The vacuum, as far as anyone knew, was really empty. He then gave it a small nonzero value in response to cosmological observations that seemed to indicate the opposite, and then later removed it again as the observations improved. A nonzero value is again in fashion due to the development of a new technique for measuring astrophysical distances using supernovae. However, none of this adjustment addresses the deeper problem. Given what we know about radioactivity and cosmic radiation, there is no reason anyone can think of why the cosmological constant should not be stupendously large—many orders of magnitude larger than the density of ordinary matter. The fact that it is so small tells us that gravity and the relativistic matter pervading the universe are fundamentally related in some mysterious way that is not yet understood, since the alternative would require a stupendous miracle.

The view of space-time as a *non-substance* with substance-like properties is neither logical nor consistent with the facts. It is instead an ideology that grew out of old battles over the validity of relativity. At its core is the belief that the symmetry of relativity *is different from all other symmetries in being absolute. It cannot be violated for any reason at any length scale, no matter how small, even in regimes where the underlying equations have never been determined.* This belief may be correct, but it is an enormous speculative leap. One can imagine moon people applying similar reasoning and chastising their brightest students for asking what the earth was made of on the ground that its roundness made the question moot. This would clearly be an injustice, since the earth is not absolutely round but only approximately so. On length scales smaller than the naked eye can easily discern from the moon, there are troublesome little details such as the Mount Everest. Advances in observation technology would eventually vindicate the students, at least the ones who remained defiant. It would be discovered that the earth is not perfectly round, and moreover is approximately round for the reason that the rocks from which it is made become plastic at the high pressures found underground, so that large objects on the surface slowly sink.

Despite its having become embedded in the discipline, the idea of absolute symmetry makes no sense. Symmetries are caused by things, not the cause of things. If relativity is always true, then there has to be an underlying reason. Attempts to evade this problem inevitably result in contradictions. Thus if we try to write down relativistic equations describing the spectroscopy of the vacuum, we discover that the equations are mathematical nonsense unless either relativity or gauge invariance, an equally important symmetry, is postulated to fail at extremely short distances. No workable fix to this problem has ever been discovered.

Thus the innocent observation that the vacuum of space is empty is not innocent at all, but is instead compelling evidence that light and gravity are linked and probably both collective in nature. Real light, like real quantum-mechanical sound, differs from its idealized

Newtonian counterpart in containing energy even when it is stone cold. According to the principle of relativity, this energy should have generated mass, and this, in turn, should have generated gravity. We have no idea why it does not, so we deal with the problem the way a government might, namely by simply declaring empty space not to gravitate. It also demonstrates the severity of the problem, for one does not resort to such desperate measures when there are reasonable alternatives. The desire to explain away the gravity paradox microscopically is also the motivation for the invention of *supersymmetry*, a mathematical construction that assigns a special complementary partner to every known elementary particle.[2] Were a superpartner ever discovered in nature, the hope for a reductionist explanation for the emptiness of space might be rekindled, but this has not happened, at least not yet. The unsubstantiated belief of Lorentz and Poincare day was ether, or more precisely the naive version of ether that preceded relativity. The unsubstantiated belief of our day is relativity itself. It would be perfectly in character to reexamine the facts, toss them over in his mind, and conclude that beloved principle of relativity was not fundamental at all but emergent—a collective property of the matter constituting space-time that becomes increasingly exact at long length scales but fails at short ones. This is a different idea from Lorentz, Poincare, Heaviside original one but something fully compatible with it logically, and even more exciting and potentially important. It would mean that the fabric of space-time was not simply the stage on which life played out but an organizational phenomenon, and that there might be something beyond.

John Bell suggestion is like going back to relativity as it was before Einstein, when people like Lorentz and Poincare thought that there was an *aether* - a preferred frame of reference - but that our measuring instruments were distorted by motion in such a way that we could not detect motion through the *aether*. Now, in that way we can imagine that there is a preferred frame of reference, and in this preferred frame of reference things do go faster than light. But then in other frames of reference when they seem to go not only faster than light but backwards in time, that is an optical illusion. Behind the apparent Lorentz invariance of the phenomena, there is a deeper level which is not Lorentz invariant. This is not sufficiently emphasized in textbooks, that the pre-Einstein position of Lorentz and Poincare, Larmor and Fitzgerald was perfectly coherent, and is not inconsistent with relativity theory. The idea that there is an *aether*, and these Fitzgerald contractions and Larmor dilations occur, and that as a result the instruments do not detect motion through the *aether* - that is a perfectly coherent point of view. I think that the idea of the *aether* should be taught to students as a pedagogical device, because I find that there are lots of problems which are solved more easily by imagining the existence of an *aether*. The reason me must to go back to the idea of an *aether* here is because in these EPR experiments there is the suggestion that behind the scenes something is going faster than light. Now, if all Lorentz frames are equivalent, that also means that things can go backward in time. That is the big problem. And so it's precisely to avoid these that I want to say there is a real causal sequence which is defined in the *aether*. Now the mystery is, as with Lorentz and Poincare, that this aether does not show up at the observational level. It is as if there is some kind of conspiracy, that something is going on behind the scenes which is not allowed to appear on the scenes. That's extremely uncomfortable.

Model of the ESP Phenomena Signaling Models

In producing a physical theory of psi, we need to decide whether we are demanding a new paradigm of physics or merely tinkering with the current one. It is natural to start off by trying the second (less radical) approach, and there are many reviews of 'tinkering' models. However, the danger is that one will end up grafting so many extra bits onto the old paradigm (like adding epicycles to the Ptolemaic model of the Solar System) that it becomes hopelessly complicated. There is also the problem of testability: there are actually many models for psi and, by adding enough bits to the standard paradigm, one can doubtless explain anything. However, a crucial requirement of a scientific theory is that it should be falsifiable and, as emphasized by many theories are inadequate in this respect. Nevertheless, it will be useful to start off by reviewing less radical approaches, since some aspects of these may still feature in the new paradigm. The discussion below, groups theories of psi into three general categories: field or *signalling* models, quantum models and higher-dimensional models.

Many theories of ESP can be viewed as 'signalling' models, in the sense that they involve the transmission of information or energy via some sort of particle or field (these concepts being linked in modern physics). Often the field involved is already part of the current paradigm. This includes, for example, explaining ESP in terms of electromagnetic waves or neutrinos It also includes explaining PK in terms of electrostatic forces Models which explain precognition in terms of tachyons or advanced waves might also be regarded as being within the current paradigm, even though they involve rather exotic aspects of it. Even more extreme are models which adopt the spirit of the current paradigm but invoke particles like *psitrons* or *ESP waves* with the specific purpose of explaining ESP. All these approaches might be regarded as tinkering with the current paradigm. Generally speaking, the experimental evidence indicates that ESP can occur at great distances and does not decline with distance. These findings do not fit well with most hypotheses that physical energies mediate the transmission of extrasensory information. Indeed, the information transmission model may itself be erroneous. However, as discussed below, even if signalling models cannot work in four dimensions, they may still be viable in higher dimensions, since the viewer and the viewed may become contiguous in the higher-dimensional space. This is a crucial feature of my own proposal.

There are also many theories which invoke some form of biophysical field, even though the status of such fields is questionable from a physicist's perspective. Mesmer's early ideas on animal magnetism and *vitalistic* fluids might be included in this category. Unfortunately, none of these approaches has gained general acceptance among *paraphysicists* and all of them have been criticized on the grounds that they are ad hoc and *unfalsifiable*. On the other hand, the link with biology is important and reflects the growing interaction between physicists and biologists in orthodox science. It also raises the issue of whether psi is involved in some forms of complementary medicine and in reincarnation cases, and whether it is a feature of mind alone or life in general.

Quantum theory — which for present purposes we regard as part of the current paradigm — provides at least some scope for an interaction of consciousness with the physical world. It also completely demolishes our normal concepts of physical reality, so it is not surprising that some physicists have seen in its weirdness some hope for explaining psi. Indeed, E. H. Walker (1984a) has argued that only quantum theory can explain ESP. The most concrete realization of the quantum approach is 'observational theory', according to which consciousness not only collapses the wave-function but also introduces a bias in how it collapses. In this picture all psi is interpreted as a form of PK which results from the process of observation itself (i.e. there must be some kind of feedback). For example, clairvoyance is supposed to occur because the mind collapses the wave-function of the target to the state reported. This process can even explain retro-PK), since it is assumed that a quantum system is not in a well-defined state until it has been observed. Another feature of observational theory is that the brain is regarded as being akin to an REG. Thus an ordinary act of will occurs because the mind influences its own brain, and telepathy occurs because the mind of the agent influences the brain of the percipient. Of course, there is still the question of how consciousness collapses the wave-function (Stapp, 1993). One possibility is to modify the Schrodinger equation in some way Observational theory has the virtue that it can make quantitative predictions. For example, one can estimate the magnitude of PK effects on the basis that the brain has a certain information output and the results seem comparable with what is observed in macro-PK effects. On the other hand, observational theory also faces serious criticisms. One can object on the grounds that psi sometimes occurs without any feedback. For example,(Beloff) has pointed out that there are pure clairvoyance experiments in which only a computer ever knows the target. One can also question the logical coherence of explaining psi merely on the grounds that one observes it and there are alternative models for retro-PK Finally, David Bohm (1986) has cautioned that the conditions in which quantum mechanics apply (low temperatures or microscopic scales) are very different from those relevant to the brain.

Nevertheless, many physicists back some form of quantum approach Some proposals exploit the non-locality of quantum theory, as illustrated by the famous EPR paradox. An atom decays into two particles, which go in opposite directions and must have opposite (but undetermined) spins. If at some later time we measure the spin of one of the particles, the other particle is forced instantaneously into the opposite spin-state, even though this violates causality. This non-locality effect is described as 'entanglement' and) tried to explain this in terms of hidden variables, which he invoked as a way of rendering quantum theory deterministic. Experiments later confirmed the non-locality prediction (Aspect) and thereby excluded at least some models with hidden variables (though not Bohm's). Indeed, John Bell, who played a key role in developing these arguments and was much influenced by Bohm's ideas, compared the non-locality property to telepathy. Einstein made the same comparison, although he intended it to be disparaging!

Although quantum entanglement has now been experimentally verified up to the scale of macroscopic molecules, it must be stressed that it is not supposed to allow the transmission of information (i.e. no signal is involved). For example, attributing remote viewing to this effect would violate orthodox quantum theory. Theorists have reacted to this in two ways. Some have tried to identify what changes are necessary in quantum theory in order to allow non-local signalling. More generally, Jack Sarfatti (1998) has argued that signal non-locality could still be allowed in some form of 'post-quantum' theory which incorporates consciousness. He regards signal locality as the micro-quantum limit of a more general non-equilibrium macro-quantum theory The relationship between micro and macro quantum theory is then similar to that between special and general relativity, with consciousness being intrinsically non-local and analogous to curvature. His model involves non-linear corrections to the Schrodinger equation and may permit retrocausal and remote viewing effects Others accept that there is no signalling but invoke a 'generalized' quantum theory which exploits entanglement to explain psi acausally. This is also a feature of the model of pragmatic information, which interprets psi effects as meaningful non-local correlations between a person and a target system. This model may account for many of the observed features of psi, including the difficulty of replicating psi under laboratory conditions). It may also be relevant to homeopathy. Radin has argued that entanglement is fundamental to ESP. This is because he regards elementary-particle entanglement, bio-entanglement (neurons), sentient-entanglement (consciousness), psycho-entanglement (psi) and socio-entanglement (global mind) as forming a continuum, even though there is an explanatory gap (and sceptics might argue an evidential gap) after the second step. If the Universe were fully entangled like this, he argues that we might occasionally feel connected to others at a distance and know things without use of the ordinary senses. This idea goes back to Bohm, who argued that there is a holistic element in the Universe, with everything being interconnected in an implicate order which underlies the explicit structure of the world:-The essential features of the implicate order are that the whole Universe is in some way enfolded in everything and that each thing is enfolded in the whole. This implicit order is perhaps mediated by ESP. Most mainstream physicists regard such ideas as an unwarranted extension of standard quantum theory, but one clearly needs some sort of extension if one wants to incorporate mind into physics.

There are various other quantum-related approaches to explaining ESP. Some of these exploit the effects of 'zero point fluctuations vacuum energy. This is a perfectly respectable physical notion, so it is not surprising that some people have tried to relate this to the traditional metaphysical idea that there is some all-pervasive energy field which connects living beings (eg. *chi, qi, prana, elan vital*). Indeed, Puthoff views the zero-point-energy sea as a blank matrix upon which coherent patterns can be written. These correspond to particles and fields at one extreme and living structures at the other, so some connection with psi is not excluded. A related proposal is that the radiation associated with zero-point-energy might be identified with *subtle energy fields* These allegedly involve some form of unified energy of such low intensity that it cannot be measured directly In the electromagnetic context, this idea was introduced to describe the quantum potential and maybe relevant to Bohm's, (Bohm, 1986) implicate order. Although these ideas might be regarded as being on the fringe of the standard paradigm, the recent discovery that 70% of the mass of the Universe is in the form of 'dark energy'—most naturally identified with vacuum energy — is stimulating interest in this sort of approach. For example, Sarfatti (2006) has a model which associates both consciousness and dark energy with the effects of vacuum fluctuations, although he does not explicitly identify them.

It should be cautioned that the literature in this area comes from both expert physicists and non-specialist *popularizers,* so it is important to discriminate between. Although quantum theory is likely to play some role in a physical model for psi, my own view is that a full explanation of psi will require a paradigm which goes beyond standard quantum theory. Of course, nobody understands quantum theory anyway, so claiming that it explains psi is not particularly elucidating—it just replaces one mystery with another one. Also, many of the above proposals already deviate from standard quantum theory, so this raises the question of how radical a deviation is required in order to qualify as a new paradigm. In my view, most of those mentioned above are insufficiently radical and one needs a new approach —perhaps of the kind envisaged by Bohm—that can explain both psi and quantum theory. One also suspects that the new paradigm will incorporate the idea of *retrocausality* discussed earlier, since proposed tests of this all involve some form of EPR effect (Cramer)

In this paragraphwe present theoretical model for the emission of the Heaviside type wave model for the remote viewing phenomena. In our ealier papers we developed the quantum model for the emission of the brain waves. In order to put forward the classical theory of the brain waves we quantize the brain wave field. In the model we assume (i) the brain is the thermal source in local equilibrium with temperature *T*. The spectrum of the brain waves is quantized according to formula

$$E = \hbar\omega \tag{25.1}$$

where E is the photon energy in eV, $\hbar$=Planck constant, $\omega = 2\pi\nu, \nu$ -is the frequency in Hz. (iii). The number of photons emitted by brain is proportional to the $(\text{amplitude})^2$ as for classical waves. The energies of the photons are the maximum values of energies of waves For the emission of black body brain waves we propose the well know formula for the black body radiation.

In thermodynamics we consider Planck type formula for probability P (E) dE for the emission of the particle (photons as well as particles with m≠0) with energy (E,E+dE) by the source with temperature T is equal to:

$$P(E)dE = BE^2 e^{(-E/kT)} dE \tag{25.2}$$

where B= normalization constant, E=total energy of the particle, k = Boltzmann constant=1.3 x 10^{-23} J K^{-1}. K is for Kelvin degree. However in many applications in nuclear and elementary particles physics kT is recalculated in units of energy. To that aim we note that for 1K, kT is equal k1K = K x 1. 3 10^{-23} J x K^{-1}= 1.3 10^{-23} Joule or kT for 1K is equivalent to 1.3 10^{-23} Joule= 1.3 10^{-23} /(1.6 10^{-19}) eV = 0.8 10^{-4} eV. Eventually we obtain 1K= 0.8 10^{-4} eV, and 1eV= 1.2 10^4 K

$$\frac{dN}{dE} = BE_{\max}^2 e^{(-\frac{E_{\max}}{T})} \tag{25.3}$$

where, B is the normalization constant, T is the temperature of the brain thermal source in eV. The function $\frac{dN}{dE}$ describes the energy spectrum of the emitted brain photons.

For the ESP phenomena we propose the emission by the source with temperature $T=10^{-15}$ eV the charged particle with charge = electron charge and mass < 10^{-15} eV which propagate in aether with velocities greater than the light velocity and emits the Heaviside type waves.

O. Heaviside claimed never to have been seduced by increasing mass with velocity:

I will not go so far as to say,that the view which is popular now, that "mass" is due to electromagnetic inertia, is a mere Will o' the Wisp. I will however say that the light it gives is somewhat feeble and uncertain, and that it eludes or evades distinct localisation. The mere ***idea,*** *that electromagnetic inertia* ***might*** *account for "mass", occurred to me in my earliest work on moving charges, but it seemed so vague and unsupported by evidence, that I set it on one side. It explains too much, and it does not explain enough.*

One curious feature of the predicted mass variation with speed is the infinity that results at the speed of light. This is a result we interpret today to mean that nothing with mass can travel fast as light (in a vacuum). Light can travel (obviously) as fast as light because photons are massless. Searle studied the electromagnetic effects of moving charges for decades do not accepted this conclusion, but Heaviside did not. Scattered all through his books, in fact, *i* analyses on charges moving /aster-than-light (with the earliest dating from 1888[4]

Heaviside would have none of Searle's caution. In 1898 he wrote a reply[7] in which he attempted to refute Searle's position. I don't really find Heaviside's prose arguments (there is not a single line of mathematics) very convincing—here is his opening "proof that Searle was wrong:

The argument ... seems to be that since the calculated energy of a charged body is infinite ... at the speed of light, and since this energy must be derived from an external source, an infinite amount of work must be done, that is, an infinite resistance will be experienced. There is a fallacy here. One easy way of disproving the argument ... is to use not one, but two bodies, one positively and the other negatively charged to the same degree. Then the infinity disappears, and there you are, with finite energy when moving at the speed of light.

The appearance of infinites in his work, an event that makes most analysts stop to ponder at * hat they might mean, hardly ever caused Heaviside to do anything but dip his pen into the well for more ink to write why there was no need for concern. For example (writing on a problem different from moving charges), he once declared, When mathematicians come to an infinity they are nonplused, and hedge around it .... We must not be afraid of infinity.

Heaviside was absolutely right about his claims for hyperlight motion // the medium is something other than a vacuum, such as water. Then the speed of light is less than it is in a vacuum, charged particles can exceed the speed of light, and, in fact, Heaviside's conical, electromagnetic shock wave is observed today we call it Cherenkov radiation after the Russian physicist P. A. Czerenkow who exhaustively studied it experimentally in the 1930s, although Madame Skłodowska Curie apparently the first to notice, in 1910, this radiation effect in radium solutions (but she did not appreciate its true origin).

Oliver Heaviside, showed that a point charge q in steady rectilinear motion along the axis of z, at a speed u, less than c, was associated with the potential

$$V = \frac{q}{4\pi\varepsilon\left[z^2 + \gamma^{-2}(x^2 + y^2)\right]^{\frac{1}{2}}} \tag{25.4}$$

where γ^{-2} is a fraction ranging from 1 to 0, as u increases from 0 to c. Here it is to be understood that V is the potential at the point x, y, z when the origin is at the charge, so that V accompanies q in its motion. It is further to be understood that the electric force **E** is derived from the potential in the manner specified by

$$E = -\nabla V \tag{25.5}$$

The question now to be considered is what occurs when u is greater than c. Are the formulae still valid? We can see immediately that some reservations are necessary, even though no change of formula may be required. For γ^{-2} is now negative; and V, and also E and H are made imaginary when

$$z^2 < (\left(\frac{u}{c}\right)^2 - 1)\left(x^2 + y^2\right) = Cot^2\theta\left(x^2 + y^2\right) \tag{25.6}$$

This means that V is real inside the two cones to right and left of the moving charge whose angles are 2θ , equation (25.6), but unreal in the intermediate region outside the cones.

But next, seeing that disturbances are propagated only at speed *c*, whilst the charge q moves at the greater speed *u*, the locus of the spherical waves sent out by the charge as it moves along forms the left conical surface only. So we must reject the right cone altogether, if we are considering a charge brought from rest up to speed u.

So far is rejection without change. But closer consideration will make it probable, if not certain, that a change in the formula is wanted as well. For, assuming that equation (25.1) is correct when κ• is negative, provided we keep to real values, it still belongs to both cones. Now it was standardised so to make the total displacement leaving the charge be q. This was with u<v, when the displacement emanated in all directions. As we employ now the same formula, the same property should hold good, keeping to the real values, however. But V is symmetrical. At corresponding points in the two cones V is the same. So the displacement leaving q for the right cone can be only and similarly for the left cone. The practical meaning is that if we reject the right cone, and still have the charge at the apex of the left cone represented by q, we must double the right side of equation (25.1). That is to say,

$$V = \frac{q}{2\pi\varepsilon[z^2 - s^2(x^2 + y^2)]^{\frac{1}{2}}} \tag{25.7}$$

should be the proper solution. Similarly, the right members of the formula for E and H, true when u<c, must be doubled when u>c.

At any point P inside the cone, we have

$$V = \frac{qTan(\theta)}{2\pi\varepsilon[z^2Tan^2(\theta) - (x^2 + y^2]^{\frac{1}{2}}} \tag{25.8}$$

So V is a minimum on the axis, and increases to infinity on the cone. Outside the cone V is zero. Deriving the electric force by equation (25.2), it will be found that **E** is radial, and is directed towards the charge. This is inside the cone. Its size is

$$E = \frac{qrTan(\theta)}{2\pi\varepsilon l^2} \tag{25.9}$$

at distance r from the apex, where l is the geometrical mean of the distances of the point P from the surfaces of the cone. The conical surface is the seat of a sheet of displacement away from the apex. This follows because V suddenly drops to zero outside the cone.

The probability of the emission of particle with mass m by the source with temperature T is governed by Boltzmann factor, $P\left(\frac{m}{T}\right)$

$$\sim P\left(\frac{m}{T}\right) \approx Exp[-\frac{m}{T}] \tag{25.10}$$

From formula (25.10) we conclude that the source with temperature T emits the particles of mass m of the order of T (temperature and mass in energy units). The theoretical spectra are presented in Figure 1. The emitted particle with velocity u propagate through the aether and emits the electric field E (25.9). The electric field, *Heaviside quanton* interacts with subject brain and create the ESP phenomena,

The comparison of the Heaviside wave, E, with experimental data of ESP.

1. In accordance with points 7 and 8 the Heaviside particle with mass m reaches the subject brain at once (velocity u >>v and generates electric field E in brain cells
2. The field E is concentrated in very narrow angle θ . In a sense the subject (source) sees the receiver subject
3. The field E does not depend on the distance of the subjects

Hypothesis, Elusive Heaviside Particles H^{+}, H^{-}

We argue that the new particle called *Heaviside (H)* particles are the carriers of ESP phenomena. The *Heavisid*e particle are proposed in the frame of the new law *electroweak baryogenesis*

The word *baryogenesis* refers to the generation of *baryons* (particles such as *protons and neutrons*) and leptons (particles such as electrons and neutrinos) out of energy states. But in physics, a process can be reversed. Tippler assumed that the process worked in reverse-baryons and leptons(p+ e) annihilate and produced the pair ($\nu + \overline{\nu}$) In our hypothesis of EPS phenomena instead of the neutrinos the new particles (antiparticles), for the moment *nonobservable* are produced in the reaction, formula (8)

$$e^{-} + p \rightarrow H^{+} + H^{-} + B \tag{25.11}$$

In formula (3.11) B is the hydrogen binding energy B $\approx$ 1 GeV, the masses of, $H^{+,-}$ are equal $m_{H^{+,-}} = 10^{-15} eV$, the charge of Heaviside particle $q_{+,-}$ = charge of electron. If *the weak baryogenesis* exists the *quantons* $H^{+,-}$ with very high energy will cause the subjects brain atoms to recoil and so the recoiling atoms would leave the tracks The tracks can be observed in brain matter with transmission electron microscope (TEM) The ESP phenomena are rooted in subnuclear physics. First of all supported by the results of the Oliver Heaviside

(1850-1925) we conclude that special relativity is not in opposition to the existence of the particles with *finite mass and velocities greater than light velocity.* The spark of ESP phenomenon in "source" subject is created by the emission of the new particle- antiparticle *Heaviside quantons*, which consists of *Heaviside particles* with mass of the order of 10^{-15} eV, which propagate with velocity greater than the light velocity. The recombination of the Heaviside pair generate san additional hydrogen atom in the brain medium of the receiving subject.

Chapter 26

LANGEVIN TWINS AND BRAIN ACTIVITY

Consciousness is nonlocal that is everywhere in nonlocal space and intrinsically entangled with all potential information stored in wave functions. Consciousness triggers collapse of the wave function and is thus the source of embodied waking consciousness. There is theoretical possibility that consciousness in nonlocal space is linked to-or serves as the basis for-the electromagnetic field connected to the nervous system and the brain

Pim van Lommel, 2012

*Langevin twins*The following remarks focus on Whitehead's early conception of space-time elaborated between the first edition of *An Enquiry Concerning the Principle of Natural Knowledge* (1919) and *The Principle of Relativity* (1922). They assume from the start that this conception provides an overarching scheme for organizing concrete spatio-temporal perspectives on the basis of idealized chronogeometrical constituents (event-particles, point tracks, etc., reached through the method of extensive abstraction), thus bridging the gap between the world of modern physics and that of ordinary sensed experience. "Space-time," in this particular context, is not to be confused with Minkowski's four-dimensional "absolute world" as a general geometrical framework for the formulation of the laws of physics. If one of Whitehead's aims is indeed to show that the interplay of spatio-temporal perspectives derived from sensed experience can be reconciled (or at least articulated) with the Minkowskian geometry of space-time describing the interrelations of such objects as point-events and world-lines, this aim cannot be achieved unless one avoids the "fallacy of misplaced concreteness" and refrains from giving undue ontological weight to what is merely the result of abstractive operations in the process of formalization[2]. Whitehead's space-time, in this sense, is a genuine philosophical construct. It includes its own relation to the space-time of physicists, while remaining distinct from it, just as the notion of simultaneity, as a relation that must somehow be directly given in perception (within the experience of "duration"), is sharply distinguished from "simultaneity" in Einstein's sense.

In *Duration and Simultaneity* (1922), Bergson describes Whitehead's *The Concept of Nature* (1920) as "an admirable book," "one of the most profound ever written on the philosophy nature[4]." This enthusiastic comment occurs in a passage devoted to the idea of an irreducible "advance of Nature" which any conception of nature should take into account[5]. But in his discussion of relativity theory, and precisely because of his unconditional attachment to the idea of temporal becoming, Bergson shows a much stronger distrust towards the geometrization of spatio-temporal relations achieved by the theory of relativity. In fact, he hardly develops a personal account of space-time. "Space-time," in his book, always refers to Minkowskian space-time, a formalization of spatial and temporal relations which, according to him, epitomizes the "spatialization of time" that lies at the very heart of modern science ever since time was defined as a parameter for describing motion. However, Bergson's attack on the timeless view of four-dimensional space-time, along with his critical examination of certain crucial points in the standard interpretation of special relativity (dislocation of simultaneity, dilatation of time, etc.), should not obscure the fact that he consistently claims to be a devoted relativist, although of a peculiar breed. His aim, as it be clear from anyone who takes the pain to actually read his book (rather than learn about its alleged "blunders" from secondary sources), is to defend Einstein against the relativist *metaphysics* that obscure the philosophical core of his new physics. One of his favourite and most notorious targets is the paradox of the twins (also known as "Langevin's" paradox or the "clock paradox").

Although the details of his deconstruction of the paradox would require a specific contribution, it is easy to sketch an outline of the main motivations behind it. This will allow us to raise a more general problem which is in fact common to Whitehead and Bergson: how is it possible to give an account of the paradoxes of relativistic time (and more particularly of the paradox of the twins) in terms of a philosophical reconstruction which does not take Minkowskian space-time for granted? As we shall see, Whitehead and Bergson take different stands on this issue, while sharing a common concern. Both insist on addressing the paradox in a way that does not lose touch with the concrete experience of the passage of nature. Yet, in contrast with Bergson, Whitehead does not attempt to deny the physical consequences of the paradox. Quite the contrary, he provides an illuminating analysis of its actual mechanism in terms of simultaneity relations, without leaving the domain of the special theory of relativity. We shall try to show in what sense his conclusions may contribute to a deeper understanding of time (and space-time) within the larger context of the "advance of Nature."

The aging twins: where is the paradox anyway?Let us start first with a little reminder. Although the original idea is present in Einstein's notorious 1905 paper on the electrodynamics of moving bodies published in the Annalen der Physik, it is Langevin who gave the paradox its most popular form, involving as it does two brothers, one of them travelling back and forth at a far distance in a rocket ship with a velocity close to that of light, the other remaining on Earth. The Langevin version of the paradox was presented to a philosophical audience in 1911 at an international symposium held in Bologna, and later at the French Philosophical Society. Bergson meditated it for ten years before publishing his book on relativity. There, he calls the brothers "Pierre" and "Paul," not in order to make the story sound more realistic or to give it a typically French flavour, but as a matter of convenience: we should remember the sedentary twin as Pierre, the one who stays still, like a rock (une pierre)... Langevin's point, at any rate, is that when the travelling twin returns to Earth and the twins are once again united in the same place, the clocks in the rocket ship show that 2 years have elapsed, while similarly constructed Earth-bound clocks show that 200 years have elapsed. These figures (and their ratio) are quite arbitrary. They vary according to the speeds attained by the rocket: the fastest the travelling speed, the more dramatic the time lag between the two clocks. This results from a mechanical application of the Lorentz transformations governing the transition from one inertial frame to another. The important point, however, is that it does not concern clocks only, but any periodical process whatsoever —that is, any process that may be timed by a clock. Thus, Langevin tells us, the Earth-twin has been aging more than his twin brother during the "same" amount of time. Or, to avoid any unnecessary paradoxical formulation at this stage, the Earth-twin has been aging more than his twin brother during a temporal separation they measure in different ways. This thought experiment is really a fancy way of presenting spatio-temporal properties that are directly derivable from the Lorentz transformations. It was originally designed to underline the need for a radical reformation of our common-sense concepts of space and time. As any standard introduction to special relativity readily acknowledges, it was confirmed later by experimental tests involving giant accelerators, the decay of elementary particles in the atmosphere or the transportation of clocks in rocket planes around the globe. This, however, need not really concern us here.

The philosophical issue is one of interpretation. It raises the question of how one should understand a statement such as: "time runs slower in the rocket ship." What does it mean for time to run slower (or faster, depending on the point of view)? The temptation here would be to dismiss the whole issue as a mere quibble on words. Time does not run faster or slower: it is only measured in different ways, according to the system of reference adopted for carrying out the operations of measurement. Very well then: for the Earth-clock to run faster only means that the measure of the interval between two limiting events (departure and arrival) yields higher numbers from the Earth's perspective than similar measures taken from the rocket ship's perspective. And this, one may add, trivially follows from the fundamental equations of the theory. Unfortunately, this will not do. What we need is a philosophical interpretation of this state of affairs. For Langevin's formulation of the paradox involves more than the mere discrepancy of clock readings: it seems to bear on the issue of perceived or lived time, in so far as it stages two "percipient events,'' as Whitehead would put it. Moreover, even if this aspect of the story is put aside and the twins are replaced by mere mechanical time-keeping devices, the paradox survives in an even stronger form as sheer astonishment yields to genuine philosophical perplexity. The true paradox lies in the fact that if only *relative* motions are taken into consideration (as the special theory invites us to do), which set of clock is regarded as travelling becomes an arbitrary matter, and thus the slowing down effect would seem to be perfectly reciprocal. Each clock having equal reasons to be considered at rest, they should eventually display equal time lags: the paradox is that the anticipated retardation effect vanishes as soon as it is symmetrically attributed to both set of clocks, and yet the theory tells us that only one twin must have aged less. An equivalent formulation would be the following: the paradox is that in order to take the slowing down of clocks seriously (as the theory encourages us to do), one must give up the symmetry clause and thus deny that there is any paradox at all. For if the effect (the differential aging of the twins) is real and thus genuinely asymmetrical, then there is no paradox, merely an unusual outcome of the theory, well confirmed by various experiments. On the other hand, if there is no paradox, then what remains to be explained is the asymmetry between the relatively moving twins. There lies the conceptual difficulty behind Langevin's eerie tale. The paradox is a puzzle about relative motion and its place in relativity theory.

BERGSON'S TWINS: SEPARATE BUT EQUAL

Bergson was very much aware of the logical structure of the paradox and he consistently emphasized the reciprocal character of temporal distortions. He is indeed famous (that is, almost universally blamed) for *not* accepting the theoretical consequences of Langevin's thought experiment. His strategy is clearly that of a dissolver or a "debunker" rather than a solver. Arguing on the grounds of a perfect *symmetry* between the two twins, he writes in the first appendix to *Duration and Simultaneity*:

> "We are dealing, in short, with two systems, S and S', which nothing prevents us from assuming to be identical: and one sees that since Peter and Paul regard themselves, each respectively, as a system of reference and are thereby immobilized, their situations are interchangeable."

This comes as the conclusion of a passage which explicitly relies on the supposed reciprocity of the motions involved in the situation:

> "If we stand outside the theory of relativity, we can quite readily conceive of an absolutely motionless individual, Peter, at point A, next to an absolutely motionless cannon; we can also conceive of an individual, Paul, inside a projectile launched far out from Peter, moving in a straight line with absolutely uniform motion toward point B and then returning, still in a straight line with absolutely uniform motion, to point A. But, from the standpoint of the theory of relativity, there is no longer any absolute motion or absolute immobility. The first of the two phases just mentioned then becomes simply an increasing distance apart between Peter and Paul; and the second, a decreasing one. We can therefore say, at will, that Paul is moving away from and then drawing closer to Peter, or that Peter is moving away from and then drawing closer to Paul. If I am with Peter, who then chooses himself as system of reference, it is Peter who is motionless; and I explain the gradual widening of the gap by saying that the projectile is leaving the cannon, and the gradual narrowing, by saying that the projectile is returning to it. If I am with Paul, now adopting himself as system of reference, I explain the widening and narrowing by saying that it is Peter, together with the cannon and the earth, who is leaving and then returning to Paul. The symmetry is perfect. We are dealing, in short, with two systems, S and S', which nothing prevents us from assuming to be identical: and one sees that since Peter and Paul regard themselves, each respectively, as a system of reference and are thereby immobilized, their situations are interchangeable."

Bergson's assumptions at first reinforce, or give more intensity to the paradox: for obviously we cannot fudge the issue by saying that the twins think the same thing of each other. A decision must be made: the twins cannot both be younger when they are reunited in the same place and in the same reference frame. That would be a plain logical contradiction and if special relativity entailed such a thing, it would itself be inconsistent. But certainly Bergson does not mean to say that. As a consequence, the only way out is to dissolve the paradox as a mere illusion involved in its very premises. The proper way to consider the issue is to acknowledge that there is genuinely no such thing as a slowing down of temporal processes in the rocket-ship (or on the Earth, for that matter). As Bergson puts it:

"The Paul who has impressions is a Paul who has lived in *the interval, and the Paul who has lived in the interval is a Paul who was interchangeable with Peter at every moment, who occupied a time identical with Peter's and aged just as much as Peter.*"

Hence, according to Bergson, the paradox simply does not arise. We merely thought there was one. We were under the illusion of a paradox. On returning to Earth, the travelling twin will have aged "just as much as" the sedentary twin. The conclusion is unambiguous: it is a *non sequitur* for many physicists and philosophers who find the statement so baffling that they would rather dismiss the whole book as a piece of philosophical sophistry than spend efforts trying to understand its inner logic. Even the more charitable readings of Bergson generally agree that his book provides a distorted view of the nature of the theory of relativity, confusing as it may seem the principle of relativity at work in Einstein's theory with the classical (Cartesian) idea of the relativity of motion.

Granted, Bergson was mistaken in considering that there is no way, within the conceptual structure of special relativity, to differentiate between the twins. Somewhere in the book, he observes in a footnote that, strictly speaking, the conditions set by Langevin exceed the confines of special relativity. The special theory should deal with frames of reference translated relatively to each other with uniform speeds. Here, however, it is clear that accelerations are involved (in the minimal sense that the traveller twin changes the direction of its journey at mid-point, not to mention the departure and arrival which also imply accelerations or at least changes in the direction of the speed vector). Of course this does not really bother Bergson, for he claims to have found a solution which amounts to dissolving the paradox, or make it vanish, as he says. Yet his solution is not valid, for it overlooks the fact that the twins are simply not interchangeable. For even if one were to consider that the motion of the twins is relative in its various stages and hence strictly reciprocal, this does not entail that the relativistic effects are reciprocal, as a simple consideration shows. It is indeed an obvious fact that the twins are separated and reunited. No one will dispute this. But what it means is that one of them (at least) must be attached to (at least) two successive frames of reference, and *this,* of course, is not reciprocal. Never mind which of the twins is *really* moving: the crucial point is that one of them must experience three events: departure, turnaround and return, whereas the other experiences only two, departure and return (or, if one prefers, separation and reunion, which are neutral as to the motions involved). Whatever one may think of the implications of the relativity of motion, the twins do not share the same space-time history (or "life history," in Whitehead's terms) and thus cannot be regarded as interchangeable clones. This simple fact is what Bergson seems to have overlooked in his search for a philosophical dissolution of the twins paradox. Surely, it is not doing much justice to Bergson to quote his lines without giving them further context. Maybe a few remarks regarding his general strategy are in order here, if only to avoid the usual misunderstandings.

Measured time and real time. The reservations one may have concerning Bergson's general line of argument should not obscure what is most interesting in the detail of his analysis. To put the matter broadly, what is at stake in the convoluted passages devoted to Langevin's paradox is the incidence of different time-systems on the actual passing of time, or if one prefers, the relation between measured time intervals and perceived or lived time intervals. This question, it must be emphasized, does not boil down to the problem of making room for "psychological time" alongside "physical time" within the framework of physical science, as Einstein believed. In this regard, the discussion between Einstein and Bergson that took place in Paris in 1922 is a rather saddening illustration of what it means for physicists and philosophers to talk at cross-purpose. Einstein never understood Bergson's point, which was quite simple: the only reason the measured, calculated, inferred time of distant events is called "time" is that it can be related to an immediate, intuitive experience of simultaneity and succession. "Real time" provides the "flesh" that is lacking from the formalization of spatio-temporal perspectives in terms of algebraic transformations. Whitehead, as we shall see, shares the same premises, although he is cautious enough not to make any direct use of the notion of "real time," considered in isolation from its articulation with an experience of "duration" or "simultaneity" that is intimately linked with spatial extension.

The dilatation of time as an effect of kinetic perspective. Bergson repeatedly claims that the so-called "dilatation" effect attributed to relatively moving clocks is a mere illusion — or to be more accurate, a "perspective effect," as opposed to an actual retardation or slowing down of real time. According to the "dilatation" effect, time intervals measured at a distance by a moving reference system appear to be longer than they would be if they were measured at the same place, by a system at rest. This is again trivially implied by the "k factor" displayed by the Lorentz equations. Einstein notoriously established this point (which was missed by both Lorentz and Poincaré) in his 1905 article, and later in his 1917 introduction to relativity theory, with a famous thought experiment involving lightning bolts and a train moving along an embankment. Bergson has no claims against the equations that capture this unsettling result. Yet he be lieves and consistently argues that the dilatation or slowing down effect is indeed a mere appearance or, better, the outcome of some sort of kinetic perspective characterized by an essential reciprocity. This point recurs in many forms throughout his book. It is an essential piece of Bergson's argumentative strategy concerning the twins paradox and it is necessary to give a full quotation here:

> "We have stated but cannot repeat often enough: in the theory of relativity, the slowing of clocks is only as real as the shrinking of objects by distance. The shrinking of receding objects is the way the eye takes note of their recession. The slowing of the clock in motion is the way the theory of relativity takes note of its motion: this slowing measures the difference, or 'distance', in speed between the speed of the moving system to which the clock is attached and the speed, assumed to be zero, of the system of reference, which is motionless by definition; it is a perspective effect. Just as upon reaching a distant object we see it in its true size and then see shrink the object we have just left, so the physicist, going from system to system, will always find the same real time in the systems in which he installs himself and which, by that very fact, he immobilizes, but will always, in keeping with the perspective of relativity, have to attribute more or less slowed times to the systems which he vacates, and which, by that very fact, he sets in motion at greater or lesser speeds. Now, if I reasoned about someone far away, whom distance has reduced to the size of a midget, as about a genuine midget, that is, as about someone who is and acts like a midget, I would end in paradoxes or contradictions; as a midget, he is 'phantasmal', the shortening of his figure being only an indication of his distance from me. No less paradoxical will be the results if I give to the wholly ideal, phantasmal clock that tells time in the moving system in the perspective of relativity, the status of a real clock telling this time to a real observer. My distantly-removed individuals are real enough and, as real, retain their size; it is as midgets that they are phantasmal. In the same way, the clocks that shift with respect to

motionless me are indeed real clocks; but insofar as they are real, they run like mine and tell the same time as mine; it is insofar as they run more slowly and tell a different time that they become phantasmal, like people who have degenerated into midgets."

The notion that time dilatation is a mere illusion is sometimes challenged on the grounds that the dilatation effect is an observable phenomena, and that as such it should be considered as no less real than time measures carried out at the same place (or "along" the world line of some system). There is no reason, the argument goes, to make any difference between measures, as long as their results can be empirically ascertained. These objections, however, miss the point. One may well concede that what is observable is real in some sense, but the problem remains untouched: Bergson believes that if motion makes any difference, then all measures cannot be real in the same sense. More precisely, there must be a difference between "direct" measures (involving a single clock at rest relative to the two events defining some temporal interval) and "indirect" measures (involving two separate clocks). The notion of "perspective effect" neatly captures this difference. By the same token, it draws our attention to the structural symmetry of the situation. No one seriously contests that the dilatation effect is reciprocal, but when it comes to assessing its exact nature, optical metaphors involving the distortions of visual perspective are more illuminating than the dubious image of a mysterious elastic time-substance subjected to various degrees of stress.

Proper time Versus Coordinate Time

This leads to another point. It is important, when reading Bergson, to refrain from adopting an overly psychological understanding of his arguments. Bergson misleadingly speaks of "lived time" (as opposed to measured, coordinate time), or "living and conscious observers" (as opposed to frames of reference with their corresponding classes of coordinate systems). But there is a formal counterpart to each of these experiential notions, and it is the job of *Duration and Simultaneity* to unravel the formal construction that spells out the original relativistic intuition in order to get a grasp on the actual points of connexion through which the theory manages to tap in the experience of real time, in spite of its tendency to frame it according to spatializing procedures. An essential aspect of Bergson's analyses consists in emphasizing the necessity of not losing sight of *intrinsic* magnitudes within the context of relativity theory. Intrinsic magnitudes are generally characterized as invariant magnitudes. Such are so-called *proper* time and lengths. Of course, Bergson has his own interpretation of invariance, and it is true that the idea of situation is more important in his eyes than the coordinating function of frames of reference in providing equivalent descriptions. As a consequence, he cannot satisfy himself with holding a four-dimensional invariant magnitude such as proper time as a definition of real time. Yet, when one translates relativistic results in terms of "proper" rather than "improper" time, many of Bergson's claims become not only harmless but quite relevant to a proper understanding of relativity. For one thing, when two systems of reference are set in relative *uniform* motion, the so-called "dilatation of time" *is* reciprocal, while the proper times, as measured by the clocks "at rest" in each system, are identical. This shows that nothing peculiar happens to the clocks: their beat is unaffected by relative motion, they remain in perfect synchrony. It remains to be seen whether this conclusion can be extended to the case of accelerated frames. Bergson's criticisms, at any rate, are a welcome warning against the kind of uncritical realism that attaches to notions such as dilated time, multiple time rhythms, and so on.

THE "TOPOLOGICAL" INVARIANCE OF THE TIME INTERVAL

Now as far as the twins paradox is concerned, things are of course a little more complex since, as we have said, more than two frames of reference are involved. It is important however to remember that Bergson is chiefly concerned with the time interval as such, not with the particular measures attributed to it by various coordinate systems. The fifth chapter of *Duration and Simultaneity* devoted to "light figures" clearly emphasizes this point: we ought to make sense of the fact that two events in time (two events separated by a "time-like" interval, in the Minkowskian parlance), although they may be causally connected by different means, implying different kinds of motions, still present themselves as one and the same time interval —or more exactly, one and the same space-time interval, explored and measured along different routes. When we say that the Earth-twin has aged 200 years "during the time" it took the traveller to reach the remote star and return back home, or conversely when we say that "meanwhile" the traveller-twin has aged 2 years, we speak improperly, for obviously the *time* interval, being measured differently by each, can count as "the same" only in a Pickwickian sense. And yet we feel there must be something common to both experiences, an underlying process of nature behind the endpoints materialized by the two meeting events (the twin's departure and his return). Is the intuition of a real time interval underlying its various measures, compatible with relativity theory? Bergson thinks it is. Accordingly, he never challenges Einstein on his own ground, nor criticizes relativity as a *physical* theory (there is nothing resembling a neo-Lorentzian temptation here). What is at stake is the philosophical interpretation that physicists and philosophers alike are prone to associate with it, specially when it comes to assessing the reality of time. Bergson's challenge is to show that there must be, within relativity's own terms, some kind of topological invariant underlying discrepant measures of time. The spatio-temporal interval between two events is frame- invariant in any case. But when it comes to two causally connectible events, this interval is time-like, meaning that it reduces to proper time if one chooses the appropriate reference frame. Thus, referring to an underlying (real) time interval as a shorthand for the spatio-temporal interval does not seem wholly off the mark even if one assumes, as Bergson does, that this time interval does not have any intrinsic measure (contrary to the spatio-temporal interval which is a metrical invariant). In this sense it is perfectly acceptable, and philosophically to the point, to speak of the twins as relating to the *same duration* in different ways, provided that such a duration is not conflated with any of its various measures.

Now, to come back to Bergson's treatment of the paradox, it seems that the intuition developed in relation to the basic situation of two frames in relative (uniform, rectilinear) motion, the intuition of the dilatation of time as a mere "perspective effect," is simply carried over to the case of the paradoxical twins. This transfer from the basic situation to the more complex one is arguable. My feeling is that it is legitimate even if one acknowledges a fundamental asymmetry in the spatio-temporal histories of the twins. There are, indeed, reciprocal dilatation effects involved in the various parts of the story, if the frames are considered two at a time. The only problem is that they are not relevant as far as the differential aging of the twins is concerned: the reciprocity of the slowing down of clocks in the simple case loses its explanatory virtues when applied to the more complex case. The reason for the temporal gap between the twins must lie elsewhere. Whitehead's strategic move, as we shall see, consists in shifting our attention to the dislocation of simultaneity, rather than the dilatation of time. When all is said and done, however, the erroneous claim that the twin traveller ages "just the same" as his brother should not obscure Bergson's original problem, which I believe remains intact even if one accepts the reality of the phenomenon anticipated by Einstein and Langevin. The problem, to put it bluntly, goes somewhat like this: in what sense is it legitimate to describe the twins as living in *different times*? More generally: what kind of metaphysical conclusion (if any) regarding the fundamental unity of real time do the unsettling results of relativity theory impose on us?

THE UNITY OF TIME AND THE INTUITION OF LOCAL TIME

A final reminder, which should close this series of apologetic remarks. Bergson never considered restoring an absolute, universal Time in Newton's sense ("equably flowing...") or even in Lorentz' sense (an absolute or "true" time attached to a reference frame at rest in the ether). Bergson claims to be a full-blooded relativist, a partisan of "radical relativity," determined to defend its philosophical core against the unwarranted generalizations of self-made metaphysicians. What he wants to show is that besides the different time- systems attached to various artificial (and in that sense arbitrary) measurement procedures, there is a common time, a unique time, "real time," on which any measure must ultimately rely if it is to be considered as a measure *of time*. Needless to say, such a time has nothing to do with the piling up of successive hyperplanes of simultaneity (or instantaneous "nows" corresponding to states of the universe at an instant) determined by some preferred foliation of space-time. Such a figuration of the "advance of Nature" as the homogeneous flow of a global time extended over the whole of space, unrolling in its uniform course as some kind of cosmic ribbon, implies the very notion of spatialized time which Bergson strives to overcome through the method of intuition. The problem with the standards accounts of relativity theory is that they generally provide a mere relativization of the traditional picture, which naturally leads to the notion of space-time as a timeless "block universe" where nothing unfolds anymore and the flow of time seems frozen, reduced as it is to the parameterization of time-like curves or "world lines." Bergson is naturally critical about this new Parmenidean stance and the naivety with which it claims to derive immediate metaphysical interpretations from the geometrical properties of a diagrammatic portrayal of the situation. But regardless of the inadequacy of the "block universe" as a model of relativistic physics, what is striking is that it is essentially homogeneous with the traditional (Newtonian) view in one important respect: time is once more equated with space, or with an additional *dimension* of space. It plays the role of a "fourth dimension of space," although in the Minkowskian case the articulation of these dimensions is more intimate and intricate than ever due to the particular structure and symmetries of the Lorentz group.

To put it bluntly, one may say that Bergson is not so much interested in universal time (as an alternative to relativistic times) as he is in the unity of time. This unity is to be contrasted with the multiple durations suggested by the dislocation of simultaneity and the slowing down of clocks. It is required if one wants to make sense of nature as a connected whole. The very possibility of a philosophical cosmology depends on it. What is the ground of this unity of time according to Bergson is a matter I do not wish to consider now. The fundamental point is that the unity of time has nothing to do with the global time-frame of traditional physics (and metaphysics), nor with the relativized times of relativity. The unity of time must be reached *locally,* in relation to the fundamental experience of simultaneity, an experience of the "now" which may well be inseparable from an experience of the "here" of a situated observer. Hence, the local nature of time is ultimately related to the way co-present events or observers experience the passage of nature. As we shall see, Whitehead's own definition of simultaneity in terms of "presentational immediacy," with its distinctive dialectics of "duration" and "cogredience," concurs with Bergson on certain fundamental points, while distinguishing itself by its special emphasis on the local construction of *space-time,* rather than time alone.

I hope it is clear from the preceding remarks that Bergson's unsuccessful attempt at dissolving the twins paradox is not wholly misguided, or at least that it makes perfect sense not only in relation to Bergson's own philosophical agenda, but within the wider arena of conceptual problems raised by the theory of relativity. Let us now turn to Whitehead and see if his solution fares any better.

WHITEHEAD'S SOLUTION: A BLIND SPOT IN SPACE-TIME

To our my knowledge, there is only one instance of Whitehead explicitly dealing with the paradox, which may suggest that he considered it of secondary philosophical importance. It is likely that in his eyes the paradox was a mere curiosity, a specimen of the kind of relativistic puzzles that a correct use of spatio-temporal diagrams could easily dispose of. One might as well say that the paradox fudged deeper issues which he had addressed at length in his books, dealing with them from scratch in terms of a new logic of spatio-temporal perspectives based on the method of extensive abstraction. Nevertheless, a puzzle must be solved, and the twin paradox offered a nice occasion for Whitehead to apply the tools he had developed in his previous writings. Whitehead's complete analysis and resolution of the paradox can be found in the *Aristotelian Society Supplementary, Volume III,* published in 1923 under the title "The Problem of Simultaneity: Is There a Paradox in the Principle of Relativity in Regard to the Relation of Time Measured to Time Lived?" The articles gathered under this heading are the outcome of a discussion or "symposium" involving a philosopher, H. Wildon Carr and a mathematician, R. A. Sampson, in addition to Whitehead himself.

It must be emphasised that Whitehead's contribution is one of the earliest (if not the earliest) presentation of the paradox that casts it entirely within the conceptual frame of the *special* theory of relativity. Contrary to what is still claimed by many authors, the paradox can be given a solution in principle in the terms of the special theory alone, provided it is purged from unnecessary realistic elements (such as engine firings, inertial effects or accelerations other than mere change of direction, not to mention gravitational fields). Whitehead's simple, elegant solution hardly comes as a revelation today: it can be found in many (though not all) introductory textbooks to special relativity. To my knowledge, however, its earliest formulation is to be found in Whitehead's contribution to the 1923 symposium. At first sight, Whitehead's reformulation of the paradox hardly qualifies as a "solution": it looks more like an analysis requiring no particular philosophical clarification or justification, admitting a trivially equivalent formulation in terms of Minkowski diagrams. Maybe this feeling is well grounded and Whitehead's solution is no solution after all. But then again, if Langevin's paradox is no paradox at all, it does not need to be solved, but only given a proper analysis. At any rate, this should not distract us from understanding in what sense Whitehead's analysis manages to shed some light on "the Relation of Time Measured to Time Lived."

The phrasing of the conference program is itself a clear indication that Bergson's earlier attempt is at the back of everyone's mind. But the main difference with Bergson is that Whitehead aims at a solution, or at least a complete clarification, not a dissolution. For one thing, he never disputes the final asymmetry between the twins. This acknowledgement of the effectiveness of the paradox is coherent with his commitment to the multiplicity of time-systems as an expression of the creativeadvance of nature: "This passage is not adequately expressed by any one time-system. The whole set of time-systems [...] expresses the totality of those properties of the creative advance of nature which are capable of being rendered explicit in thought." Obviously, the existence of alternative time-systems is no objection against the unity of real time. Time-systems are already fairly elaborate (and hence abstract) constructs evolving from a more primitive experience of simultaneity. So there is no point in trying to establish the uniqueness of time-systems: time-systems or temporal perspectives cannot be one, as a matter of definition. If there were only one time- system, it would not be a temporal perspective. Bergson himself never doubted that time-systems, attached as they are to the experience of inertial frames, would yield the shifting simultaneity relations described by the Lorentz transformations and captured in the space-time formalism of Minkowski. But his idea in *Duration and Simultaneity* was to root these artefacts in the local, purely temporal experience of simultaneity which he describes, fundamentally, as a simultaneity of *flows* rather than moments. Through the local connecting links provided by neighbouring virtual consciousnesses disseminated throughout nature as a whole, two experiences, however far apart, always "have some part in common": step by step, they can be "reunited in a single experience, unfolding in a single duration," a duration which is, so to speak, unaffected by spatial separation. Whitehead's own account of simultaneity must be contrasted with this view. His intuition is that one should refrain from folding back the whole spatio-temporal structure unto the experience of a situated consciousness. One should make room for real time at an intermediary level involving a genuine experience of *distance* as such. "Duration," as he understands it, is an intrinsically *spatio*-temporal concept; simultaneity is indissolubly local *and* distant. But this spatial component also provides the key to an adequate solution of the paradox. Let us see how this works. Whitehead's solution, as mentioned earlier, can be given an immediate translation in Minkowskian terms. However, in its 1923 formulation it is not couched in the usual four-dimensional idiom. It does not involve proper times measured along the respective world lines of the twins. Whitehead offers instead a solution that deliberately ignores the intrinsic formulation in terms of absolute or invariant magnitudes.

Quite consistently, the emphasis is on the discrepancy between particular perspectives (or time-systems) embedded in the space-time manifold of events. Whitehead provides a detailed account of the measuring procedures on each side, as well as of the time ascriptions they involve for distant events.

The standard four-dimensional analysis, relying on the pseudo-Euclidean geometry of Minkowski space-time, would certainly save much trouble. It would imply realizing that the traveller's world line is not a geodesic —that is, not the straightest space-time path between the two given events. Whitehead does not use this trick: he sticks to the "space plus time" representation of the situation, relying on a very general argument involving the dislocation of simultaneity at the turnaround point in the travelling twin's journey, and a subsequent *blind spot* in the reckoning of distant time intervals.

For the sake of clarity, Whitehead introduces quantities, but it is not required for the understanding of his solution to the paradox.

> "Let us scrutinize more narrowly the problem of the Earth and the traveller in space. When the traveller reckons time by days, what does he count? The rotations of the Earth? Certainly not. At least, certainly not, if the traveller is to count twice 365 revolutions to the chronologer's count of two hundred times 365 revolutions. For if the traveller counts the Earth's revolutions, he presumably uses his own definition of simultaneity, and will count 3,65 revolutions of the Earth on his way out; and on his way back he will adopt another definition of simultaneity and will count another 3,65 revolutions of the Earth; in all, 7,3 revolutions of the Earth. What has happened to the remaining 72992,7 revolutions which have occurred between his departure and return? He dropped those out of account in his sudden change of space-time systems at the star, when he ceased his outward journey and commenced his return."

> "In the flurry of an instantaneous change of motion at S, the traveller dropped out of account the 72992,7 revolutions between H_1 and H_2. If he had noticed them, he would have counted them; and would then have agreed with the Earth-chronologer on his return."

Thus, the traveller's clock may appear to be out of step, but it is not slowed down in any substantial sense —only in the trivially relative sense that the rocket-time it takes the traveller to complete his journey is less than the corresponding Earth-time it takes the stay-at-home to wait for his return while he is enjoying his space-trip. The source of the confusion, as one may suspect, lies in the use of "while" in such a scenario: we tend to refer both measures to some underlying absolute temporal interval which we naturally identify with the proper time of the Earth twin. Again, days do not pass more slowly for the traveller; a day is a day and an hour lasts one hour aboard the rocket-ship as well as on Earth. It simply takes the traveller fewer days than his sibling to cover the same spatio-temporal distance, namely the interval separating his departure from his return. By taking up speed and covering more space, he has saved some time. Without even knowing it, he was relying on a structural feature of (relativistic) space-time: the speed associated with his path through spacetime was diverting motion through time into motion through space, enabling him to save time.

This very simple and yet puzzling statement is more far-reaching than it may seem. In Whitehead's view, it bears on the fundamental issue of congruence. Underlying his reflections is the following assumption: if both clocks run truly in the sense defined above, they must somehow run at the same rate throughout their whole spatio- temporal existence, as long as they are preserved from interfering forces (gravitational or otherwise). The temporal units must somehow be congruent all along, in spite of the discrepancy in the measures of total elapsed time. Those who think otherwise implicitly refer these measures to an absolute notion of the time interval, which of course makes no sense in a relativistic context. If there is no absolute time, no spatio- temporal path connecting two events qualifies a natural gauge for the passage of time between them, not even the one corresponding to the longest proper time (i.e., the geodesic defining the "spatio-temporal interval" between these events). On this point Whitehead and Bergson part ways. Remember the fourth of my "charitable remarks" above: Bergson suggests that the unity of real time lies in some underlying topological invariant which the metrical invariant merely hints at. But this is obviously of no use in the case of the twins, where one is dealing with *two* invariant measures of proper time, none of them having any special priority (as Bergson himself readily admits). Whitehead lucidly confronts this new situation. His suggestion is that just as a detour in space takes more time (all things being equal), a detour in space-time takes less time. This might sound odd, but it does not help to say that time runs slower for those who move (or accelerate, to be more precise). In fact, such statements are even more obscure than the paradox itself. There are many (spatio-)temporal routes from one event to another, but time itself need not run slower or faster for that matter.

Why is it so difficult for us to see that time does not have to be elastic in order for the twins to experience different amounts of proper time between the same pair of events? Although Whitehead does not elaborate on this point, the twins paradox illustrates the kind of confusion that arises from our continuous reliance on the ideas of absolute space and time. The source of our discomfort with the differential aging of the twins may be traced back to the absolute meaning we attach to the traveller's "mid-point" at which the shift occurs. Where, indeed, does this U-turn take place? What spatial background are we taking for granted when we try to develop a Bergsonian intuition of the continuous unfolding our twins' "contemporaneous" space-time trajectories? Similar questions would apply to the temporal aspect of the paradox. A little reflection on this particular case (and others related to the dislocation of simultaneity, such as the famous train thought-experiment devised by Einstein) may help shedding light on Whitehead's observation that

> "The paradoxes of relativity arise from the fact that we have not noticed that when we change our time-system we change the meaning of time, the meaning of space and the meaning of points of space (conceived as permanent)."

Yet, despite our difficulties in intuiting the relativistic spatio-temporal framework, there is nothing special about time itself. Again, one can claim that time flows equably without turning into an advocate of absolute Newtonian time. Whitehead touches here on a fundamental principle of his philosophy of nature, the principle of uniformity. Despite the distortions exhibited by the paradoxes, nature and the spatio- temporal structure expressing its essential relatedness, are characterized by a principle of uniformity or congruence which is in fact presupposed by any comparison of lapses of time and stretches of space. As far as time is concerned, such a claim does not reduce to the trivial assumption that time flows at the rate of one second per second. What is at stake is the commensurability of diverging timelines. Here is how Whitehead argues his case:

> "The fact of the clock running truly means that the time of one revolution of the hour hand is congruent to the time of one revolution of the Earth. But the lapse of clock time is a lapse of time according to the traveller's meaning, And this meaning differs from that for Earth time. But, though the meanings for time are different in the two cases, the lapses of time according to the different meanings are comparable as to congruence. Of course if this be denied, there can be no sense in comparing two such lapses so as to say that one lapse is 730 days and the other is 73,000 days. For in that case a day in one sense is not comparable in magnitude with a day in the other sense. But we are agreed that the days are of equal length; though there are more of them, between the departure and return of the traveller according to the Earth chronologer's meaning for time, than for the traveller's two meanings as he journeys outward and inward."

The proper times of the twins would not be comparable throughout their separation, and consequently the asymmetry found between them after they are reunited could not even be ascertainable, if it were not for this fundamental fact of congruence between their two time-systems. In Alan White's terms: Whitehead "realizes that, although the stay-at-home twin must have experienced more days (as a composite time-interval) than the star-faring sibling, any pair of days taken from both world-lines themselves must be of equal length — congruent— in some sense."

When it comes to moving from one event to another, travelling in space at great speeds appears as an advantageous solution for impatient natures: it requires less time (that is, fewer hours, fewer days) than staying at home. That is all there is to the alleged "slowing down" of clocks (or time). Time does not flow more or less rapidly, but there are *shortcuts* in space-time. The twin's space trip is a case in point: the spatio-temporal path it draws happens to yield the shortest measure of proper time. Thus, the different meanings ascribed to time do not entail any indeterminateness in the idea and experience of time itself. The blind spot phenomenon and the ensuing discrepancy in temporal measurements do not contradict the essential uniformity of the "advance of Nature," nor the congruence relations underlying the mapping of space-time by coordinate frames. As Bergson wittingly says about the twins: "We shall have to find another way of not aging."

This statement may sound obscure, but it really sums up the matter. Uniformity attaches to all perceptions in the mode of pure relatedness, but the connecting link here is the notion of simultaneity —not in Einstein's sense though, but in connection with the distinctively Whiteheadian notion of duration. Basically, the idea is that time owes its uniformity from space itself as experienced in duration. Remember that duration is described by Whitehead as a concrete "slab of nature," that is all nature present for sense awareness. But the notion of the whole of nature being disclosed in immediacy is itself the product of the "indefinitely extended projection of the 'spatial' boundaries of a single specious present." A duration is unlimited in spatial extent; simultaneity is the experience of space itself in presentational immediacy. Thus, if the issue of temporal congruence naturally leads to that of simultaneity, it is in so far as the spatial component of simultaneity yields the very possibility of congruence in a temporal context. "How am I to lay two successive revolutions of a clock hand alongside each other?," Whitehead asks. This may indeed seem a very strange request, but the spatial analogy is not wholly misguided: the periodical motions occur in space, and space is, so to speak, its own measure, since its uniform structure contains all the possibilities of congruence. These remarks may suffice to indicate how the solution works. The basic idea is that the measure of time and the issue of temporal congruence cannot be treated apart from the *spatio*-temporal expression of the passage of Nature.

There would be much to say about the way this connects to the problem of conventions, but let us save this for some other occasion and emphasize once more the insight into the fabric of space-time gained by the Whiteheadian treatment of the twins. Briefly stated: time is neither stretched nor dilated; we know this not from some inner, ineffable acquaintance with its uniformly flowing nature, but from the experience of simultaneity as a *spatio*-temporal fact of awareness. Time may be represented as a fourth dimension of space, it may be said, quite misleadingly, to be *in* space; but the truth is that it is always given *with* space as an irreducible aspect of extendedness. Time is not elastic; it flows uniformly because it is spread over space (doctrine of simultaneity) as much as it flows through space (doctrine of the passage of Nature). In return, space cuts across time, so to speak: it is responsible for the twisted, sheared aspect of space-time which the relativistic paradoxes express in their own way, but it is also what connects time with itself in a uniform manner, allowing for congruence relations to extend to the temporal realm. Bergson was ready to give up space-time altogether if only to avoid this frightful perspective: a "fibered" time uniformly flowing through the criss-crossing paths of space-time, with its innumerable creeks, shortcuts and detours. In his desperate attempt to bypass the inevitable implications of the twins paradox, he certainly missed the occasion of giving local time its proper *extension*.

Contemporary, special relativity theory is the basis for the study of the nano- and atto-world. It seems that the same holds for brain research, especially when the brain can be accelerated to high velocities. During the Apollo-Moon the astronauts brain were mowing with velocities 11 km/s. While it the minute part of light velocity, astronauts after landing on Moon reported the distortion in the brain activity, for example in description of the distance on the Moon horizon. It seems to be interesting the study of the brain activity in the circumstances of the very high velocities and applied the known results of special relativity. In paper (Marciak-Kozlowska, Kozlowski, 2012) the theoretical model for brain emission of electromagnetic waves. It was shown that the brain is the source of brain waves with temperature T- 10^{-15} eV= 10^{-10} K. In monograph (Marciak-Kozlowska,Kozlowski, 2009) it was shown that temperature of the moving thermal source with velocity v,T(v), is equal:

$$T=T(0)/(1-(v/c)^2)^{-1/2}$$

Where $T(0)$ is the temperature of the source in the source reference frame. In the following we gathered of both results and we put $T(0)$= 10^{-10} K in order to calculate the energy spectra of the brain wave in " laboratory frame of reference"

In order to put forward the classical theory of the brain waves we quantize the brain wave field. In the model (Marciak-Kozlowska, M. Kozlowski, 2012) we assume (1) the brain is the thermal source in local equilibrium with temperature T. (2) The spectrum of the brain waves is quantized according to formula

$$E=\hbar\omega \tag{26.1}$$

where E is the photon energy in eV, $\hbar$ =Planck constant, $\omega = 2\pi\nu, \nu$ -is the frequency in Hz. (3). The number of photons emitted by brain is proportional to the (amplitude)^2 as for classical waves. The energies of the photons are the maximum values of energies of waves For the emission of black body brain waves we propose the well Planck formula for the thermal radiation. In thermodynamics we consider Planck type formula for probability $P(E)$ dE for the emission of the particle (photons as well as particles with m≠0) with energy (E,E+dE) by the source with temperature T is equal to:

$$P(E)dE = BE^2 e^{(-E/kT)} dE \tag{26.2}$$

where B = normalization constant, E=total energy of the particle, k = Boltzmann constant=1.3 x 10^{-23} J K^{-1}. K is for Kelvin degree. However in many applications in nuclear and elementary particles physics kT is recalculated in units of energy. To that aim we note that for 1K, kT is equal k1K = K x 1. 3 10^{-23} J x K^{-1}= 1.3 10^{-23} Joule or kT for 1K is equivalent to 1.3 10^{-23} Joule= 1.3 10^{-23} /(1.6 10^{-19}) eV = 0.8 10^{-4} eV. Eventually we obtain 1K= 0.8 10^{-4} eV, and 1eV= 1.2 10^4 K

$$\frac{dN}{dE} = BE_{\max}^2 e^{(-\frac{E_{\max}}{T(0)})} \tag{26.3}$$

where, B is the normalization constant, T is the temperature of the brain thermal source in eV. The function $\frac{dN}{dE}$ describes the energy spectrum of the emitted brain photons. $T(0)$ is the brain waves source in the brain frame of reference. In the laboratory frame of reference in which brain is mowing with velocity v, the source temperature is equal $T(v)$

$$T(v) = \frac{T(0)}{(1-\left(\frac{v}{c}\right)^2)^{\frac{1}{2}}} \tag{26.4}$$

In the „ laboratory frame of reference" the energy spectra of brain waves are described by the formula

$$\frac{dN(v)}{dE} = BE_{\max}^2 e^{(-\frac{E_{\max}}{T(v)})} \tag{26.5}$$

In Figure 1 the spectrum of the brain waves in brain reference frame is presented. In Figure 2 we present the calculation of the energy spectrum for brain waves with brain velocities v/c=0, 0.5 and 0.9 respectively. Figure 3 represents comparison of the normalized calculated spectra formula (4)) and spectrum for v/c=0. As can be seen from formula (26.5) the brain wave source is hotter when $v \to c$. It means that the brain waves are more energetic and as the result shorter.

As the result it occurs that the emitted waves are shorter than in non-moving brain.

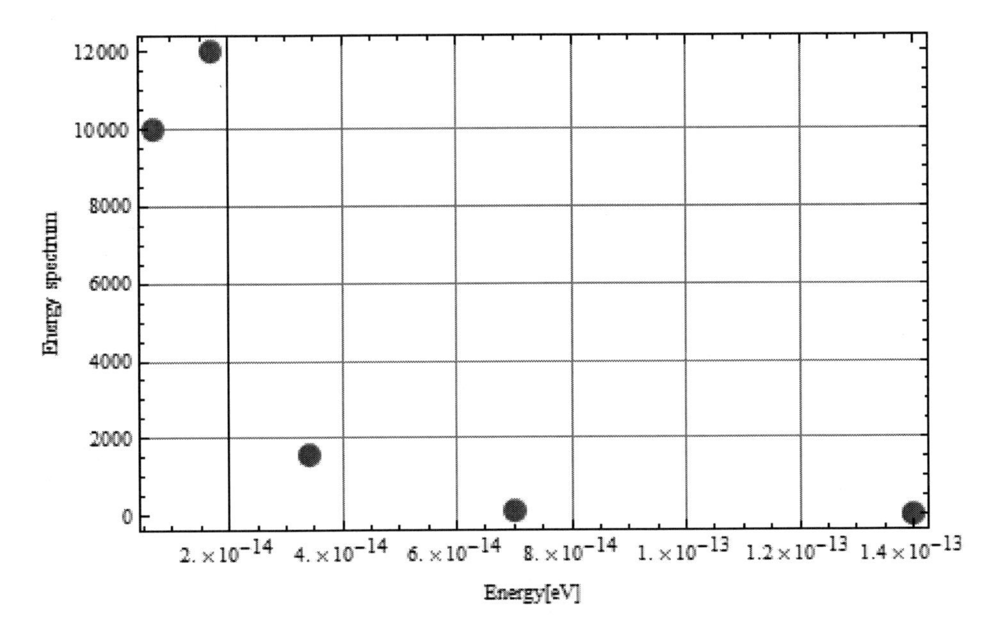

Figure 1. The energy spectrum of the brain waves in brain reference frame.

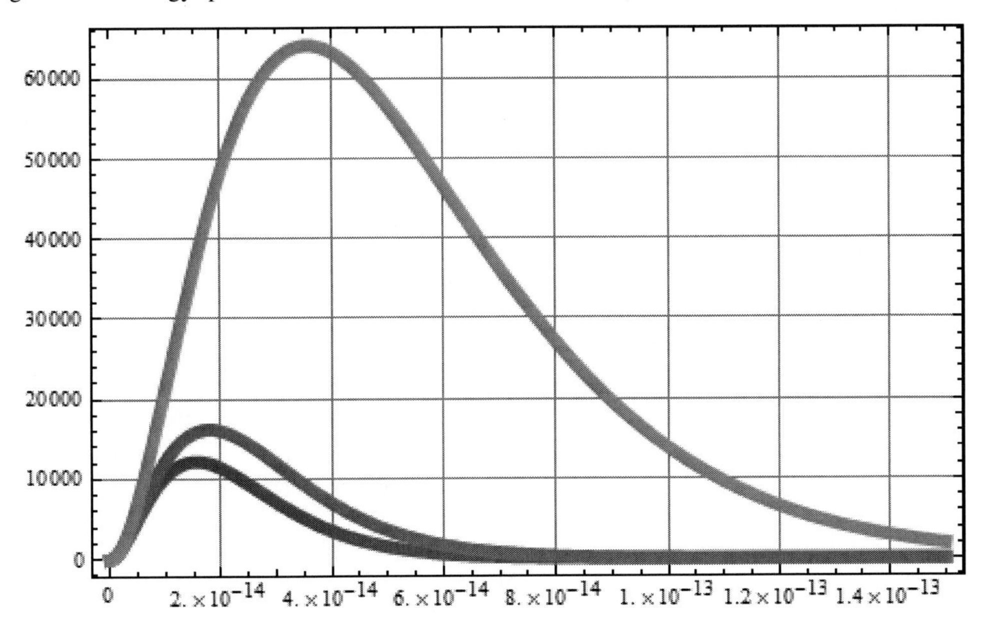

Figure 2. The calculated spectra, formula (5) of brain waves for different velocities of brain for v/c= 0 (blue), v/c=0,5 (violet), v/c=0.v=0.9 (green).

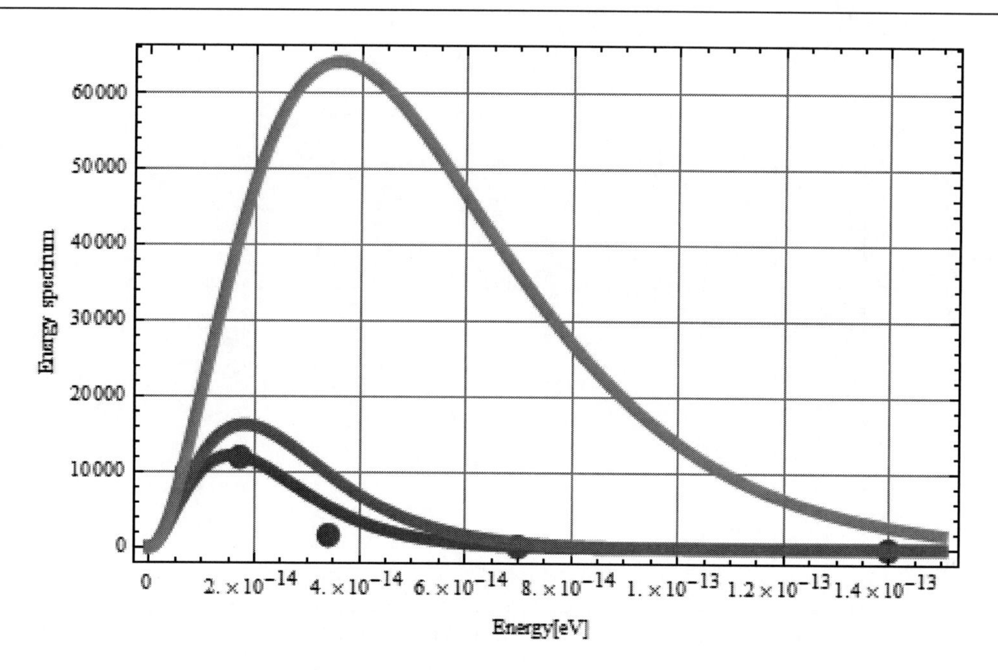

Figure 3. The calculated and measured spectra (in v=0) frame of reference.

We can formulate hypothesis that aging of the moving twin is the real phenomenon as the hotter brain source means faster metabolic processes in mobile twin body. This result solve the elusive problem of the different ages of the twins

REFERENCES

Bergson, H (1999 [1922]), *Duration and simultaneity,* transl. L. Jacobson, Manchester, Clansmen Press. (*Durée et simultanéité, in Mélanges,* A. Robinet ed., Paris, Presses Universitaires de France, 1972).

Marciak-Kozlowska J.,Kozlowski, M. (2009) *From femto-to attosecond and beyond,* NOVA Science Publishers, USA, 2009.

Whitehead A. N. (1923), "The Problem of simultaneity", *Relativity, logic, and mysticism* (Aristotelian Society Supplementary Volume III), p. 34-41.

Whitehead, A. N. (1961), *The Interpretation of science. Selected essays,* New York, Bobbs- Merrill.

AUTHORS' CONTACT INFORMATION

Janina Marciak-Kozlowska
Institute of Electron Technology
Al. Lotnik´ow 32/46, 02–668 Warsaw, Poland

Miroslaw Kozlowski
Warsaw University
Grzybowska 5/804
00-132 Warsaw, Poland
miroslawkozlowski@gmail.com

INDEX

D

E

F

G

H

I

J

K

L

M

N

O

P

Q

R

S

T

U

V

W

Y

Z